新型铋系半导体/MOFs光催化材料

郑晶静
王彩君
著

化学工业出版社

·北京·

内容简介

本书介绍了一系列铋基氧化物纳米颗粒如碘氧化铋、钒酸铋、钨酸铋与金属-有机骨架（MOFs）复合光催化材料。主要内容包括MOFs/铋基氧化物复合光催化材料的制备、性能以及载体MOFs与铋基氧化物相互作用的机理。本书重点介绍了在可见光照射下，MOFs/铋基氧化物复合光催化材料的光催化以及光电转换效率；MOFs与铋基氧化物相互作用对光生载流子的影响；探索出一条制备高活性复合光催化材料的有效途径，从而为高效光催化剂的制备提供理论依据和材料保障。

图书在版编目（CIP）数据

新型铋系半导体/MOFs光催化材料/郑晶静，王彩君著.—北京：化学工业出版社，2024.1
ISBN 978-7-122-44758-6

Ⅰ.①新… Ⅱ.①郑… ②王… Ⅲ.①铋-半导体-光催化剂 Ⅳ.① TQ426.99

中国国家版本馆CIP数据核字（2024）第001879号

责任编辑：彭爱铭　　文字编辑：王　迪　刘　璐
责任校对：刘　一　　装帧设计：孙　沁

出版发行：化学工业出版社
（北京市东城区青年湖南街13号　邮政编码100011）
印　　装：北京天宇星印刷厂
710mm×1000mm　1/16　印张 $10\frac{3}{4}$　字数169千字
2024年6月北京第1版第1次印刷

购书咨询：010-64518888　　　　　　售后服务：010-64518899
网　　址：http://www.cip.com.cn
凡购买本书，如有缺损质量问题，本社销售中心负责调换。

定　价：88.00元　　　　　　　　　　　版权所有　违者必究

前言

目前，能源短缺、环境污染等问题已然成为威胁人类生存与发展的巨大危机，如何解决这些阻碍人类社会可持续发展的障碍，成为世界各国科学家所共同面临的挑战。自从1972年日本科学家Fujishima等人发现利用TiO_2电极在光照下能够分解水产生氢气和氧气以来，许多科学家都把目光投向半导体光催化领域，这是由于一方面光催化材料具有还原作用，可以利用充足的太阳光催化，为分解水制氢提供清洁的动力能源；另一方面它还具有氧化作用，可以降解和矿化环境中的各种有机、无机污染物，解决环境污染等问题。由此可见，光催化是解决能源和环境等问题最为理想的途径之一。然而，以氧化锌（ZnO）、二氧化钛（TiO_2）等为代表的光催化纳米材料，由于具有较宽的带隙（>3eV），只能吸收日光当中的紫外光，这就严重地限制了半导体材料对太阳能的利用效率。此外，目前所报道的可见光光催化材料大多具有较高的光生电子-空穴复合率和较差的可见光吸收效率，导致它们的量子效率和可见光催化性能较低。而且在实际使用中，纳米光催化剂容易团聚，分离和循环使用比较困难。近年来，为了提高其光催化活性，已经发展了许多技术，例如：掺杂贵金属、合成量子点以及与其他半导体复合等。然而，这些光催化剂的太阳能转化效率仍然很低。因此，探索与开发高效的可见光光催化材料，促进光生电子-空穴的有效分离，提高光催化效率，仍然是目前光催化领域亟待解决的科学难题。

本书将介绍一系列铋基氧化物纳米颗粒与MOFs光催化剂结合起来制备的MOFs/铋基氧化物复合光催化材料，探索载体MOFs与铋基氧化物相互作用的机理，重点研究MOFs与铋基氧化物相互作用对光生载流子的影响，研究其在可见光照射下的光催化以及光电转换效率，探索出一条制备高活性复合光催化材料的有效途径，从而为高效光催化剂的制备提供理论依据和材料保障。

著者
2023年7月

目录

第一章　绪论　　1

  1.1　引言　　2
  1.2　光催化技术原理、影响因素、改善措施和应用　　2
  1.3　纳米孔洞金属有机骨架材料的研究背景和发展现状　　24
  1.4　铋系半导体光催化材料的研究背景和现状　　41
  1.5　主要研究内容　　60
  　　参考文献　　62

第二章　Bi_2WO_6/MIL-100（Fe）复合光催化材料制备及性能研究　　82

  2.1　引言　　83
  2.2　实验部分　　84
  2.3　结果与讨论　　86
  2.4　结论　　92
  　　参考文献　　92

第三章　BiOI/磁性树脂复合光催化材料制备及性能研究　　96

  3.1　引言　　97
  3.2　实验部分　　101
  3.3　结果与讨论　　103
  3.4　结论　　108
    参考文献　　108

第四章　UiO-66/BiOI 复合光催化材料制备及性能研究　　112

  4.1　引言　　113
  4.2　实验部分　　117
  4.3　结果与讨论　　119
  4.4　结论　　124
    参考文献　　125

第五章　Fe/W 共掺杂 $BiVO_4$/MIL-100（Fe）复合光催化材料制备及性能研究　　129

  5.1　引言　　130
  5.2　实验部分　　132
  5.3　结果与讨论　　133
  5.4　结论　　142
    参考文献　　143

第六章　纳米多孔 $FeVO_4/g-C_3N_4$ 复合光催化材料制备及性能研究　　148

  6.1　引言　　149
  6.2　实验部分　　149
  6.3　结果与讨论　　151
  6.4　结论　　154
    参考文献　　154

第七章　铁掺杂纳米多孔 $BiVO_4$/MIL-53（Fe）复合光催化材料制备及性能研究　　157

  7.1　引言　　158
  7.2　实验部分　　158
  7.3　结果与讨论　　160
  7.4　结论　　164
    参考文献　　165

第一章

绪 论

1.1 引言 2
1.2 光催化技术原理、影响因素、改善措施和应用 2
1.3 纳米孔洞金属有机骨架材料的研究背景和发展现状 24
1.4 铋系半导体光催化材料的研究背景和现状 41
1.5 主要研究内容 60
 参考文献 62

1.1 引言

自从进入 21 世纪以来，寻找一条绿色的且能够从根本上解决能源短缺、环境污染和气候变暖等全球性问题的新途径，已经成为世界各国科学家所面临的共同挑战和任务。在太阳能、风能、水能和生物能等诸多可再生能源中，太阳能具有采集方便、清洁环保、储量丰富及利用形式多样等特点，特别是其取之不尽、用之不竭的特性，可以满足人类社会可持续发展的战略需求。半导体光催化，即利用半导体的光生电子-空穴特性，将太阳能转换为电能或化学能，是太阳能最主要的利用形式之一。半导体光催化材料既可以利用太阳能分解水制氢以及将 CO_2 还原为有机低碳烷烃燃料，也可以降解环境中的各种有机、无机污染物，因此光催化是解决当今人类社会所面临的能源和环境问题最理想的途径之一。

1.2 光催化技术原理、影响因素、改善措施和应用

光催化是催化化学与光化学的交叉学科。光催化剂是指在光子激发下能起到催化作用的所有材料。最典型的天然光催化剂是叶绿素，光照下通过光合作用促进 CO_2 和水生成 O_2 和碳水化合物。最早的光催化技术是 1924 年 Baur 等研究者发现的，在光照的条件下，ZnO 能够将银盐还原为 Ag 单质，并推测此光催化反应中存在氧化还原反应[1]。1972 年，日本科学家 Fujishima 和 Honda 首次利用二氧化钛（TiO_2）在紫外光照射下把水分解为氢气和氧气[2]，成功地将太阳能转化为化学能，至此半导体光催化的研究逐渐受到关注。科研工作者经过几十年深入的研究，在光催化机理和新型光催化材料设计等方面取得了丰硕的成果[3-5]。光催化技术的优点是：反应条件温和、效率高、能耗低、无二次污染物、成本低、安全性高等。通过光催化技术，可以降解去除有机物污染物；分解水提供氢能；还原重金属；有机合成；抗菌；等等[5-9]。光催化技术已成为目前最为活跃的研究方向之一。

1.2.1 光催化技术原理

目前主要是从半导体能带理论的角度分析光催化反应机理。如图 1-1 所示：半导体的能带结构由一系列的空带和满带组成，处于顶部富载电子的是

低能价带（valence band，VB），位于底部无电子的是高能导带（conduction band，CB），导带和价带能级结构是不连续的，两者之间的间距是半导体的带隙能或禁带宽度（E_g）[10]，E_g 的大小是半导体导带位与价带位的能级差。半导体禁带宽度在 1~5eV；金属导体的禁带宽度小于 1eV；绝缘体的禁带宽度大于 5eV。

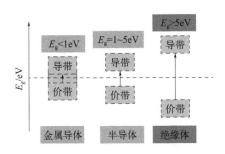

图 1-1　金属导体、半导体和绝缘体示意图

在光催化反应过程中，当半导体光催化剂受到大于或等于其禁带宽度的光子能量激发时，处于价带的电子在获得足够的能量后就会跃迁至导带上，从而在导带上形成光生电子（e^-），同时在价带中留下等量带正电的空位，称为光生空穴（h^+），此过程就生成了光生电子-空穴对（光生载流子）。具有较强还原性的光生电子和具有较强氧化性的光生空穴与材料表面吸附的反应物发生一系列氧化还原反应，从而起到产生氢气、降解有机污染物等作用[11-12]，这就是光催化反应的基本过程[13-15]。如图 1-2 所示，半导体光催化剂受到光激发，电子跃迁产生光生电子空穴，之后光生载流子发生随机运动。光生电子与空穴迁移到材料的表面并被束缚，光生电子还原氧气生成超氧自由基（·O_2^-），然后继续还原·O_2^- 生成羟基自由基（·OH），光生空穴氧化水生成·OH。氧化还原产生的·O_2^-、·OH 及光生空穴都有较强氧化还原能力，可直接与反应物反应。这个过程的反应式总结如下：

$$半导体 + h\nu \longrightarrow e^- + h^+$$

$$e^- + O_2 \longrightarrow \cdot O_2^-$$

$$h^+ + H_2O \longrightarrow \cdot OH$$

$$\cdot O_2^- / \cdot OH / h^+ + 污染物 \longrightarrow 产物$$

此外，光生电子与空穴带有相反的电荷，两者之间存在较强的库仑力，一部分光生电子与空穴不可避免地在材料内部或者表面发生复合，然后以荧光或热量的形式释放，失去氧化还原能力。决定光催化活性的关键因素之一就是半导体材料中光生电子与空穴的分离效率。因此，为提高光催化剂的活性，在实际光催化进程中，就需要采用有效手段抑制光生载流子的复合。

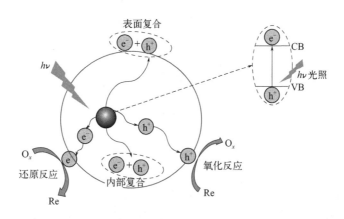

图1-2　光催化反应机理示意图

1.2.2　半导体光催化技术的影响因素

虽然光催化技术有反应条件温和、能耗低、无二次污染物等诸多优点，但是最大缺点是光催化效率不高，不能大规模实际应用。研发高效、稳定的光催化材料仍是亟待解决的核心问题。为了解决这一问题，首先对影响光催化材料催化效率的因素进行详细研究。目前，学术界普遍认为光催化剂催化效率不高的主要原因是：半导体材料本身的性质（能带结构、缺陷、晶体结构、形貌尺寸、比表面积等）；外部环境条件（光源强度、反应物浓度、反应温度等）。这些条件有的影响光生载流子的形成，有的影响其迁移和分离效率，有的影响催化材料表面反应效率。本文将主要讨论半导体材料本身性质对其光催化性能的影响，这对制备高效光催化剂至关重要。

1.2.2.1　能带及其位置

半导体的带隙宽度决定了其吸收太阳光的范围，是影响催化性能的首要因素。激发光的能量大于等于其禁带宽度，才能产生光生载流子参与后续的氧化还原反应。图 1-3 为几种常见半导体的能带结构图[16]，半导体的禁带宽度越大，吸收太阳光的范围越窄，产生的光生载流子就越少，越不利于光催化反应。例如：物理化学性质稳定、无毒、廉价的 TiO_2 因其禁带宽度（E_g≈3.2eV）较大，只能吸收约占太阳光 5% 的紫外光，从而限制其光催化性能[17]。但是，带隙宽度过小，它相应的导价带电子和空穴的氧化还原能力也会降低，光催化性能也会降低。例如，Bi_2S_3 的禁带宽度是 1.3eV，但其过低的导价带电势致使其光生载流子的氧化还原活性低，从而导致其光催化性能差[18]。半导体材料的带隙结构与光生载流子的活性有关，通常，价带电位越正，光生空穴的（h^+）氧化能力越强；导带电位越负，光生电子（e^-）的还原能力越强。但是对于特定光催化反应需满足价带电势比空穴受体电势更正和导带电势比电子受体更负。例如光解水制氢和氧：H^+/H_2 还原电位为 0eV，O_2/H_2O 氧化电位为 1.23eV，催化剂的导价带电位必须分别小于 0eV 和大于 1.23eV，并且其禁带宽度需大于 1.23eV 才能同时实现光催化制氢和氧[19]。此外导价带的氧化还原能力与半导体的禁带宽度正相关，与其对光吸收的范围负相关。所以兼顾两者的优点才能制备高效光催化剂。

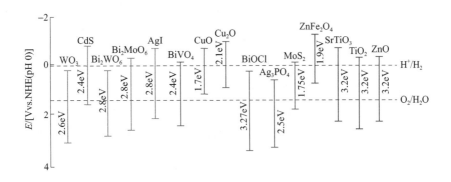

图 1-3　几种常见半导体能带结构图

1.2.2.2　晶体结构

晶体结构是影响光催化材料催化性能的一个重要因素。在合成光催化材料的过程中，由于制备条件的不同和晶面的各向异性，同一种组成的光催化材料会呈现不同晶相和结晶度以及晶相相同但晶面暴露程度不同的状态。光催化材料晶型的不同状态使其对光有不同的吸收范围、光生载流子的分离效率及表面催化效率,最终呈现不同的光催化性能。实验室中合成用于光催化的钒酸铋($BiVO_4$),其主要晶体结构是四方硅酸锆结构(tz $BiVO_4$)、单斜白钨矿结构(m $BiVO_4$)和四方白钨矿结构(ts $BiVO_4$)。研究表明 $BiVO_4$ 的光催化性能与其晶体结构密切相关。在 $BiVO_4$ 三种晶型中,m $BiVO_4$ 可见光催化活性最好。例如,Tokunaga 等[20]的研究结果表明:ts $BiVO_4$ 与 m $BiVO_4$ 具有相似的电子结构而且带隙宽度也几乎相同,在可见光的照射下,m $BiVO_4$ 具有优异的光催化活性,但是 ts $BiVO_4$ 的光催化活性几乎为零,原因是 m $BiVO_4$ 中 Bi 6s 孤对电子引起的局部结构的位错促进了光生载流子的分离,提高了光催化性能。Zhao 等[21]制备了 tz $BiVO_4$,然后通过延长水热时间将 tz $BiVO_4$ 转变为 m $BiVO_4$,m $BiVO_4$ 的光催化活性远远高于 tz $BiVO_4$。这也说明半导体晶体结构对光催化材料的催化性能起到至关重要的作用。此外,相同情况下半导体材料晶体的结晶程度也对催化性能有影响,结晶程度越高晶型越完美,越有利于光生载流子的分离与迁移,越有利于光催化反应[22]。

1.2.2.3 颗粒尺寸和形貌

催化剂的外观形貌及颗粒尺寸是影响光催化材料催化性能的重要因素。在光催化过程中,光生载流子只有迁移到催化材料表面并且没有复合时才能参与氧化还原反应,因此光催化材料颗粒尺寸越小,在光源的照射下产生的光生载流子迁移到表面的路程和时间就越短,复合的概率就越小,从而到达催化材料表面参与氧化还原反应的载流子就越多,越有利于其光催化活性的提高。另外,催化剂颗粒尺寸越小比表面积越大,参与反应与反应底物的接触面积越大,反应活性位点越多,光催化反应进行得越快,越有利于光催化性能的提高。当半导体光催化剂的颗粒尺寸小到量子级别时,颗粒的量子限域效应会引起半导体能级分立、禁带宽度变宽、光吸收范围变窄、导价带的电位变大,促进光催化反应的进行,提高其光催化性能[23]。光催化反应都是发生在光催化材料的表面,不同形貌的光催化材料具有不同的晶面暴露面、光捕获能力及反应活性位点数量等,因此光催化剂表面形貌决定其表面状况并对其光催化性能有一定的影响。从维度角度光催化材料主要分为零维、一维(纳米管、纳米线、纳米棒)、二维(纳米薄膜、纳米片)、三维(纳米花、纳米球)四类。其中一维和二维材料相比于零维材料,表面缺陷较少、不易团聚且光生载流子传递速率快，具有更好的光催化活性[24-25]。但是三维光催化材料相比于其他维度光催化材料，具有更强的光利用率、固定的反应中间产物和终产物、更好的稳定性和能回收利用等优势，因此，三维光催化材料是当前的研究热点[26]。

1.2.3 半导体光催化材料性能的改善措施

高效半导体光催化剂是光催化反应的核心，也是光催化技术大规模工业化生产应用受限的关键因素。例如，TiO_2[27]和ZnO[28]等光催化剂带隙较宽，只吸收紫外光，太阳能利用率低，同时光生载流子复合率高，致使其光催化性能较弱。CdS[29]和Ag_3PO_4[30]等带隙窄，吸收可见光，但不稳定且单组分载流子复合率高等，致使其工业化进程缓慢。因此制备高效稳定的新型可见光光催化剂是当前研究的热点。目前，半导体光催化材料性能的改善措施主要有以下几种。

1.2.3.1 高能晶面暴露

光催化材料受到光激发产生光生载流子，随后光生载流子在材料内部扩散，向表面迁移，最后在材料的表面发生氧化还原反应。表面原子的排列和配位决定了光生载流子在表面的状态、反应物分子的吸附以及产物分子的脱附，因此光催化材料的性能必然与表面原子结构有密切关系。材料表面晶面的不同取向决定表面原子的排列和配位，光催化材料表面的暴露晶面对光催化材料的活性有重大影响[31-34]，所以暴露高活性晶面是优化半导体光催化材料性能的一条有效途径。

Jiao 等人以 Au@Ag 纳米棒为晶种，用种子诱导法，成功制备了具有 {221} 和 {332} 等高指数晶面的 Ag_3PO_4 二十四面体（图 1-4）。由于 Ag_3PO_4 化学性质不稳定，无法通过透射电子显微镜（TEM）判断其表面裸露晶面，因此采用公式计算的方法推导了多面体表面的晶面指数，证实其表面存在 {221} 和 {332} 等高指数晶面，并利用密度泛函理论计算了 {221} 和 {332} 晶面的表面能，发现其具有比低指数晶面如 {100} 等更高的表面能，并且发现 {332} 晶面的导带和价带位置都略高于 {221}，从而实现了光生载流子在 Ag_3PO_4 二十四面体不同晶面间的定向传输。以上研究结果对可控构建具有高指数晶面的光催化纳米材料具有重要的参考价值和指导意义。后来 Jiao 等人继续开

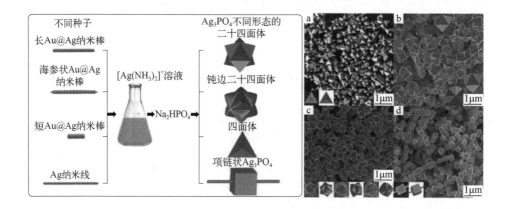

图1-4 种子诱导法制备异质光催化结构的示意图及四面体（a）二十四面体（b）具有光滑棱边的二十四面体（c）及项链状金属/Ag_3PO_4（d）SEM图

展了将 Ag_3PO_4{110} 和 {100} 面构建到同一晶体上的工作，其由 12 个 {110} 面和 6 个 {100} 面组成，且其晶面比例可理性调控，光催化研究结果表明：这些复合晶面的晶体具有比 {100} 面的 Ag_3PO_4 立方体及 {110} 面的菱形十二面体更高的光催化性能，且当 {110} 和 {100} 晶面的面积比例为 6∶4 时光催化活性最佳（图 1-5）[35]。Wang 等通过控制加入导向剂 $TiCl_3$ 的量调控 m $BiVO_4$ 晶体 {010} 晶面的暴露程度，实验结果表明：m $BiVO_4$ 光催化活性的高低依赖于 {010} 晶面的暴露程度（图 1-6）[36]。

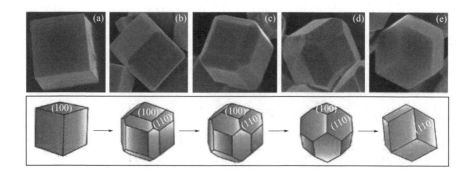

图1-5 Ag_3PO_4{110} 和 {100} 面的晶面调控 SEM 及示意图

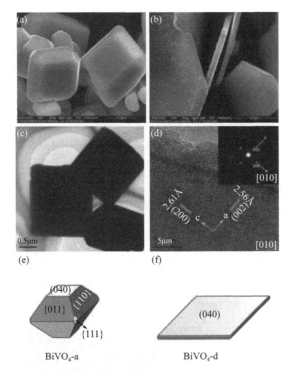

图 1-6 暴露 {010} 晶面的 m $BiVO_4$ 的 SEM 和 HRTEM 照片

1.2.3.2 掺杂

元素掺杂是指将非金属（F、Se、P、C、S 等）或金属（Fe、Co、Er、Yb、Bi 等）元素利用物理和化学方法掺入半导体的晶体结构内部。元素掺杂改变半导体化学组成，实现对其电子结构的调控，如图 1-7 所示，电负性和电子能级不同的元素掺入半导体材料中，元素能级轨道相互杂化形成新的能级就改变了原半导体的能级结构[37]。由于 $SrTiO_3$ 和 TiO_2 都是宽带隙半导体材料，无法利用太阳光中的可见光，为了实现 $SrTiO_3$/TiO_2 异质复合材料的可见光催化性能，Jiao 等人采用过渡金属离子掺杂的方法，制备了 Cr 掺杂 $SrTiO_3$/TiO_2 异质复合纳米管结构，少量铬离子掺杂可以在 $SrTiO_3$ 能带内部形成一个杂质带隙，从而有效地减小了激发 $SrTiO_3$ 所需的光源能量，实现了其在可见光下的吸收特性及光电转换性能（图 1-8）[38]。后来 Jiao 等人为了在制备 $SrTiO_3$/TiO_2 复合结构时既能够保持 TiO_2 纳米管的管状结构又能够实现复合材料对可见光的响应，同时采用控制反应动力学和铬离子掺杂的方法，首次制备了 Cr 掺杂 $SrTiO_3$ 立方体点缀的 TiO_2 纳米管阵列（图 1-9）。这种独特的异质结构有三大优点：其一，有效地减小了 $SrTiO_3$ 半导体材料的带宽，实现了对可见光的吸收与利用；其二，合理利用了 $SrTiO_3$ 与 TiO_2 的带边位置差异，提高了光生载流子的分离效率；其三，很好地保持了 TiO_2 的管状结构，利于光生电子沿着 TiO_2 纳米管快速导走。因此，这种复合结构有效地解决了 TiO_2 在光催化过程中无可见光响应和电子-空穴复合率高的两大缺陷，使其产氢速率比传统 TiO_2 纳米管高一个数量级，具有很好的创新性和应用前景[39]。

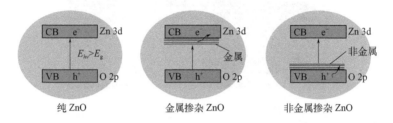

图 1-7　单体、金属掺杂、非金属掺杂 ZnO 的能带结构图

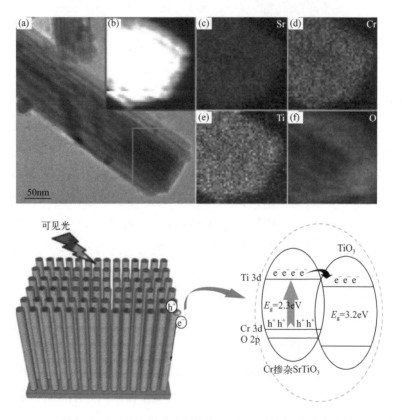

图 1-8　Cr 掺杂 $SrTiO_3/TiO_2$ 复合纳米管的 SEM、元素分布图像及电荷转移机理

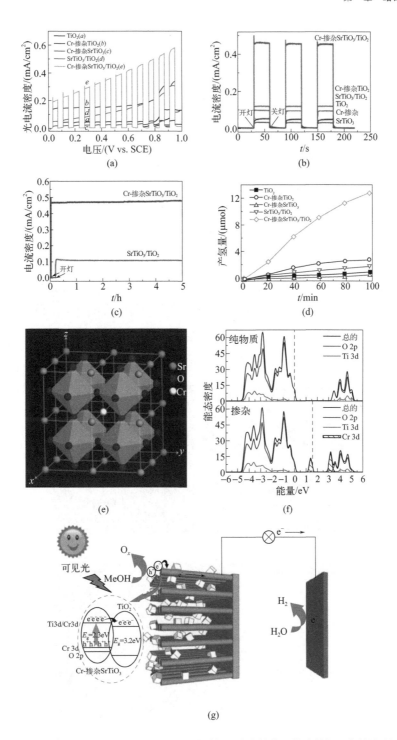

图 1-9 铬掺杂 $SrTiO_3$ 立方体/TiO_2 纳米管结构的光电性能、带隙结构及电荷转移机理

1.2.3.3 复合

过量的掺杂容易导致掺杂处成为载流子复合中心，反而降低光催化材料光生载流子迁移率。半导体材料的复合体系在光生载流子的空间迁移和提高氧化还原电势等方面有更大的优势。例如：Jiao 通过控制水热反应动力学的方法，即通过调控水热反应速度，实现了 $SrTiO_3$ 纳米颗粒在 TiO_2 纳米管顶部的定位生长，克服了在 $SrTiO_3$ 原位生长过程中，将 TiO_2 纳米管全面包覆破坏其管状结构而不利于电子定向传导的缺陷。实验证明这种方法很好地保持了 TiO_2 的管状结构，利于电子沿着 TiO_2 纳米管的管壁快速导走，从而提高了电子 - 空穴的分离效率，使其表现出更加优异的光电转换性能（图 1-10）[40]。Jiao 等人发现单质铋也具有明显的表面等离子体共振效应及光催化活性，其在 270nm 和 390nm 处有两个很强的表面等离子体共振吸收峰，吸收带边波长可以延伸至 650nm 处，由此可见金属铋在可见光区具有明显的表面等离子体共振效应，且其本身也可以作为一种光催化材料，能够利用太阳光进行分解水制氢及光电转换，并且其光催化性能可通过引入银元素形成 BiAg 合金纳米球的方法而大

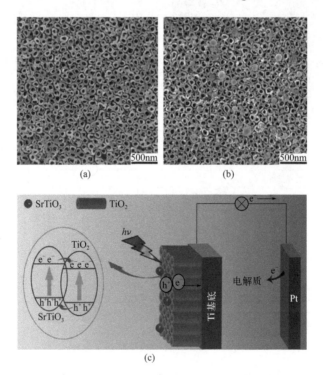

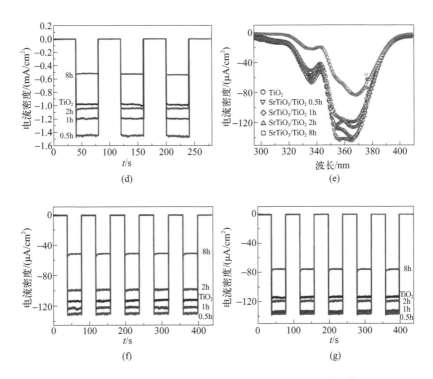

图 1-10 SrTiO$_3$ 在 TiO$_2$ 纳米管顶部定位生长的结构、性能及电荷转移机理

幅度提高，金属铋不同于一般金属而具有光催化性能的主要原因就是其具有表面等离子体共振效应，在光的激发下其电子产生跃迁并转移到费米能级更低的金属银上，从而促进其光催化及光电转换性能（图 1-11）[41]。

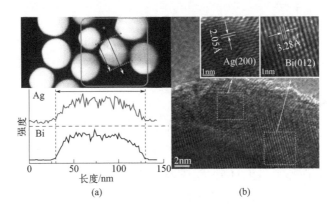

图 1-11

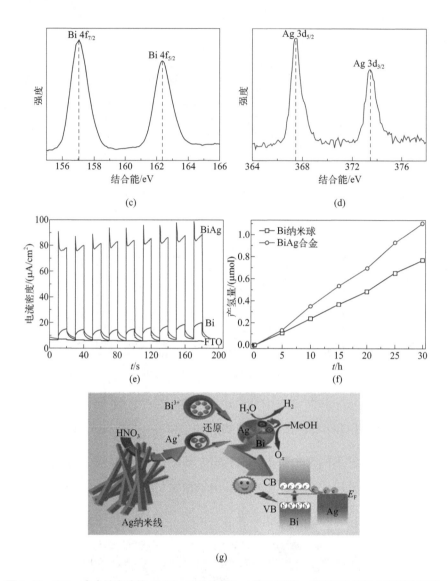

图 1-11 BiAg 合金纳米球的 SEM、TEM 图像、XPS 图谱、光催化性能及电荷转移机理

1.2.3.4 表面敏化

表面敏化是指将敏化剂沉积或吸附在禁带宽度较大的催化剂的表面,将宽带隙催化剂光吸收范围拓宽[42]。在光催化过程中,敏化剂受到光源激发后,电子由最高分子轨道(HUMO)跃迁至最低分子轨道(LUMO),转移到催化剂的导带上参与氧化还原反应,如图 1-12 所示。贵金属纳米颗粒中的自由电子在受到外界光激发的时候能够产生共振,通过调节纳米颗粒的尺寸和形貌,

这种表面等离子体共振吸收能够实现从可见光到近红外光区域的可控调节[43]。利用贵金属的这一表面等离子体共振效应，将贵金属纳米颗粒同宽禁带半导体结合，拓宽了整个光催化体系的光吸收范围，大大提高了对太阳光的利用效率[44-49]。但是由于贵金属同半导体接触时会在界面产生一个肖特基势垒，降低了贵金属被光激发产生的共振电子向半导体传导的效率[50-51]，同时贵金属价格昂贵、储量稀少，并不适合实际应用。

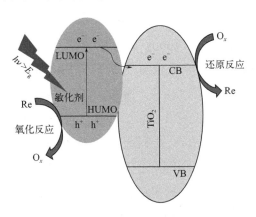

图 1-12　光敏化反应作用机制示意图

Stefan Franzen 的一项研究发现，$In_{1-x}Sn_xO_3$（ITO）薄膜同样具有表面等离子体共振效应[52]。产生这一现象的原因主要是 Sn 的大量掺杂极大地提高了 In_2O_3 中的载流子密度（10^{20}~$10^{22} cm^{-3}$），并且他提出所有的 sp 和 d 轨道电子导电的氧化物半导体都可以具备表面等离子体共振效应。这为发展新型的表面等离子体材料提供了新的途径。

后续大量的研究证明，不借助贵金属，通过增加半导体中自由电子的密度，单一的半导体同样能具备等离子体共振吸收。例如，通过掺杂的方式提高半导体中的载流子密度，在 Al 掺杂的 ZnO[53]、P 掺杂的 Si 半导体中都存在着等离子体共振吸收[54]。然而，掺杂虽然提高了半导体的光吸收范围，但是同时也会给半导体带来其他影响，并不利于半导体在光催化上的应用。例如，ZnO 中 Al 的掺杂量从 0 增加到 8% 时，对红外区的光吸收不断增强，但是能带却从 3.3eV 增加至 3.54eV，造成吸收带边蓝移，减少了对紫外光的吸收。更重要的是，异质元素的掺杂往往容易在半导体中形成载流子的复合中心而降低半导体的光催化活性[55]，这是由于大量掺杂元素的掺入并不一定形成替位掺杂，反而容易在半导体内部形成团簇[56]。因此通过元素掺杂虽然可以使半导体具有表面等离子体效应，从而具备更宽的光谱吸收范围，但由于以上原因并不利于光催化活性的提高。除了元素掺杂能够引起半导体的表面等离子体共振效应外，半导体本身存在的缺陷也能够起到增加半导体中载流子密度的作用。

Scotognella[57]和Burda[58]的研究分别发现，由于Cu缺位，$Cu_{2-x}Se$以及$Cu_{2-x}S$同样具有表面等离子体共振效应。类似地，人们发现当氧化物中存在氧空位时，也能够产生表面等离子体共振效应。在一项关于氧空位自掺杂的TiO_2的光吸收谱研究中，Murray发现氧空位自掺杂的TiO_2在红外区域光吸收增强，在3400nm处有一个表面等离子体共振吸收峰，并且由于氧空位的存在，TiO_2的吸收带边同时红移，增加了对可见光的吸收[59]，这种氧空位自掺杂的TiO_2具有很高的光催化活性，但是作者并没有对表面等离子体效应和光催化活性之间的关系做深入的研究。Alivisatos的研究发现，含氧空位的WO_{3-x}（x=0.17）同样具有表面等离子体共振吸收，在900nm处出现共振吸收峰，这与通过Mie-Gan理论计算得到的共振吸收峰位置一致[60]。而Cheng通过溶剂热法制备了氧空位自掺杂的MoO_{3-x}[61]，发现反应温度在140℃和160℃的产物的共振吸收峰分别位于680nm的可见光区和950nm的红外光区。虽然Cheng没有研究MoO_{3-x}的可见光光催化活性，但在催化分解NH_3BH_3的实验中，发现在可见光或者红外光的照射下，MoO_{3-x}催化产氢的活性比没有光照的情况下提高了数倍，而同样条件下普通的不含氧空位的MoO_3并未出现类似的效应。

氧空位作为一种自掺杂，不会在半导体中引入外来元素，关于氧空位在光催化中的作用已有众多的研究，并且人们一般认为氧空位的存在有利于提高半导体的光催化活性。例如在TiO_2中，氧空位会在TiO_2导带下方产生一个占据态能级，而这个占据态能级有效地阻止了光生电子和空穴的复合。Albert Figueras研究还发现，当氧空位的浓度升高时，占据态能级同导带杂化会缩短能带，使TiO_2能够吸收可见光，并且光催化活性增强[62]。Li通过在250~500℃氢气气氛下处理WO_3得到了含氧空位的WO_{3-x}样品，在模拟太阳光照射下其光电流密度相对于不含氧空位的WO_3样品提高了一个数量级[63]。类似地，Gan等人同样发现在引入氧空位后，In_2O_{3-n}的光电流密度大幅提高，达到了3.83mA/cm^2[64]。Wang等人利用过氧根受热剧烈分解的特点，热分解过氧化锌在氧化锌中原位引入了大量的氧空位，通过控制退火温度和气氛实现了对氧空位浓度的可控调节，利用第一性原理计算了不同氧空位浓度对 ZnO 电子结构和能带结构的影响，探讨了氧空位浓度对 ZnO 光电转换效率和光催化活性的影响，如图 1-13 所示[65]。

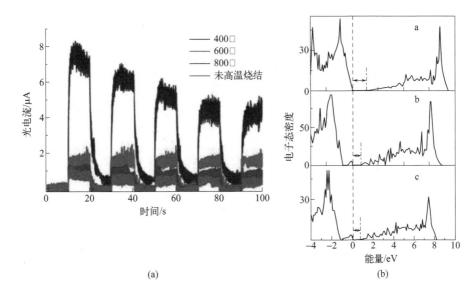

图 1-13　不同温度条件下烧结的 ZnO 样品在可见光照射下的光电响应情况（a）和电子态密度（b）

1.2.4　半导体光催化的应用领域

太阳能是取之不尽用之不竭而且廉价、清洁的能源。半导体光催化反应可以高效快捷地利用、转化太阳能，被认为是解决能源危机以及环境问题的一种环保手段。半导体光催化可应用于光催化降解污染物、光催化分解水、光催化还原二氧化碳（CO_2）、杀菌、光催化有机合成等（图 1-14）[66-72]。从反应机理上来分析：光催化降解污染物和杀菌是利用光生空穴以及产生的反应活性物质来氧化分解有机污染物的[73-74]；光催化分解水是利用光生电子与水发生还原反应产生氢气，同时光生空穴与水发生氧化反应产生氧气[75-76]；光催化还原 CO_2 是利用光生电子发生一系列还原反应[77-78]；光催化合成有机物是光生载流子的一系列氧化还原反应[79-80]。接下来将对这几个方向的应用进行深入讨论。

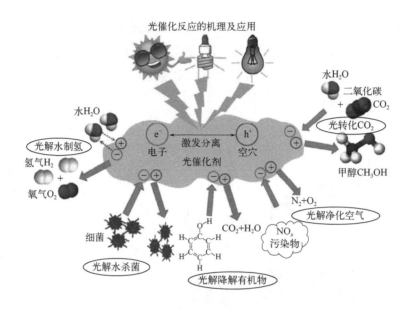

图 1-14　半导体光催化的应用领域

1.2.4.1　光催化降解污染物

半导体光催化能降解的污染物可分为：氮氧化物、硫化物、甲醛和丙酮等有害气体；亚甲基蓝、甲基蓝、罗丹明、甲基橙等液相有机染料；苯酚、甲苯、氯苯酚等芳香族化合物；砷离子、铬离子[Cr(Ⅵ)]等有毒重金属离子[81-87]。光催化材料吸收大于其禁带宽度能量的光，产生光生载流子，其中光生电子迁移到催化剂表面与吸附的 O_2 反应生成·O_2^-，同时空穴迁移到催化剂表面与水反应生成·OH。这些活性物质与污染物发生氧化反应，达到净化环境的目的。

Zhang 等研究者[88]采用水热和沉淀法制备了 Ag_3PO_4/rGO/BiOBr 复合光催化剂，实验结果表明：30min 内，Ag_3PO_4/rGO/BiOBr 对亚甲基蓝（MB）的降解率可达到 96.5%，对罗丹明 B（Rh B）、甲基橙（MO）的降解率也均在 87.5% 以上，其电子传输过程如图 1-15 所示，Ag_3PO_4 和 BiOBr 产生光生电子和空穴，而 rGO 作为催化反应的主要场所同时增加电荷迁移速率。Cheng 等通过阴离子交换工艺制备了纳米晶（NCs）Bi_2S_3/BiOCl 的复合光催化材料[89]，复合光催化材料在可见光下高效降解 2,4-二氯苯酚（2,4-DCP），其活性优于氮掺杂的 P25-TiO_2（图 1-16）。

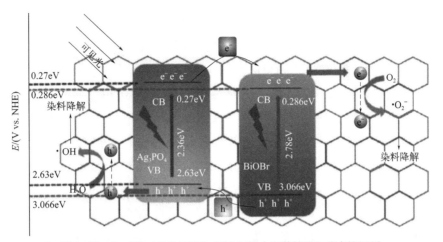

图 1-15　Ag_3PO_4/rGO/BiOBr 复合材料中电荷转移和分离的机制

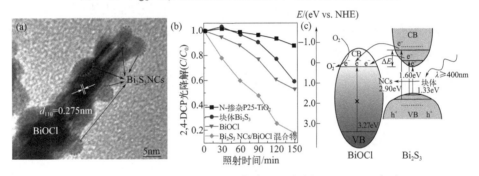

图 1-16　Bi_2S_3/BiOCl HRTEM 图（a）可见光降解 2,4-DCP（b）和 Bi_2S_3/BiOCl 载流子迁移示意图（c）

1.2.4.2　光催化分解水

光催化分解水制氢制氧是一种制取氢气的绿色环保技术，如果能够实际应用将有效缓解日益严重的能源危机和环境污染，副产物氧气也有很多用途[90-94]。从图 1-17 中可以看出：首先半导体的导带位要高于 H^+/H_2 的标准电位（0V vs. NHE），其次其价带位要低于 H_2O/O_2 的标准电位（1.23V vs. NHE），该反应才能够发生。另外半导体的光生电子和空穴分离效率要高[95-96]。反应过程如下：半

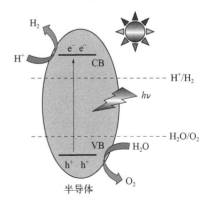

图 1-17　半导体光催化全解水的反应机理图

导体吸收光产生光生载流子，光生电子迁移到光催化剂表面然后还原吸附表面的 H^+ 生成 H_2，同时光生空穴也迁移到催化剂表面氧化吸附表面的 H_2O 生成 O_2。反应方程式：$H_2O = H_2 + 1/2 O_2$。到目前为止，如果不添加助催化剂和牺牲剂，光催化全解水的效率很低，离实际生产的标准很远，这归因于现存的半导体光催化剂的光生载流子分离效率低和催化剂表面活性位点少。为了早日实现光催化分解水产氢产氧的产业化，研究者们一直在努力。例如，Shi 等人[97]通过原位光沉积和超声剥离及组装的方法制备了 g-C_3N_4/MoS_2 纳米点和 g-C_3N_4/MoS_2 单层复合光催化剂，制备过程如图 1-18（a）所示。实验结果表明：所制备的复合光催化材料和 g-C_3N_4 纳米片相比，产氢量、活性都得到极大的提高［图 1-18（b）］。这是因为 MoS_2 纳米点和单层 MoS_2 都显著提高了光生载流子的分离和传输效率，同时两者中大量的不饱和 S 原子吸附了更多的氢离子，从而提高光催化制氢的速率和产率。Hao 等人[98]制备了大孔/介孔 TiO_2/g-C_3N_4 复合光催化材料，结果表明：TiO_2/g-C_3N_4 复合光催化材料光催化降解性能显著增强，这是因为 TiO_2 与 g-C_3N_4 形成的异质结构提高了光生载流子的分离效率，促进了光生载流子的传输（图 1-19）。

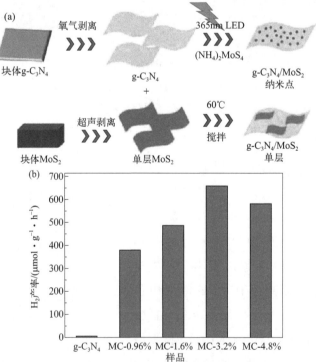

图 1-18　g-C_3N_4/MoS_2 纳米点和 g-C_3N_4/MoS_2 单层催化剂的合成示意图（a）和 g-C_3N_4/MoS_2（MC-X）的光催化产氢性能图（b）

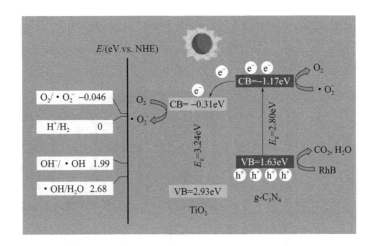

图1-19 $TiO_2/g-C_3N_4$ 复合光催化剂的降解机理

1.2.4.3 光催化杀菌

如图1-20所示为半导体光催化抗菌杀菌机理：光催化材料被光激发后，光生电子和空穴可以直接与细菌发生氧化还原反应，也可通过先与吸附在光催化材料表面的 O_2 和 H_2O 等发生反应生成 $·O_2^-$ 和 $·OH$ 反应自由基，然后这些反应自由基（具有较强的氧化能力）再直接或间接与细菌结合，致使细菌细胞结构破裂，分解细胞中的核酸等遗传物质为无害的水或二氧化碳等小分子，最后细菌完全失活[99-100]。

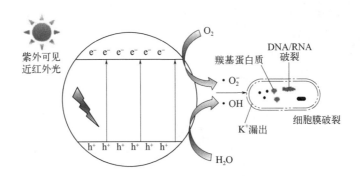

图1-20 光催化杀菌机理示意图

1988年，Matsunaga等人发现 TiO_2 光催化反应能将细胞内的辅酶A反应成二聚体辅酶A，致使细胞呼吸衰退而死亡，从而对大肠杆菌（E.coli）进行消杀（图1-21）[101]。Zeng等采用水热法制备了 $BiVO_4/Ag/g-C_3N_4$ Z型异质结，在生理盐水和污水两种环境中测试其杀灭 E.coli 的能力。实验结果表明：复合异质结相比于单独组分，光生载流子的分离效率和杀灭 E. coli 的性能都有较大提高（图 1-22）[102]。

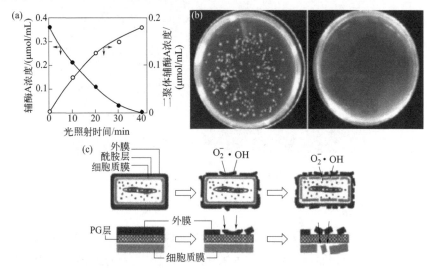

图 1-21　光催化过程中辅酶 A 和二聚体辅酶 A 的浓度随时间的变化图（a）光催化反应 4h 前后菌落图像（b）和 TiO$_2$ 薄膜杀菌过程示意图（c）

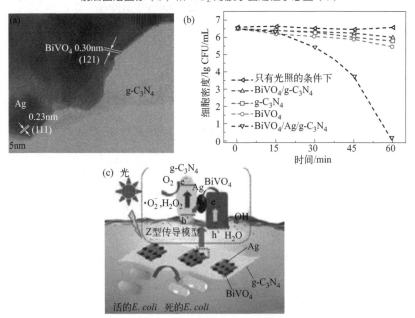

图 1-22　BiVO$_4$/Ag/g-C$_3$N$_4$ 的高分辨 TEM 图（a）不同催化剂可见光下的杀菌性能（b）和 BiVO$_4$/Ag/g-C$_3$N$_4$ 光催化杀菌机理图（c）

1.2.4.4 光催化合成有机物

受自然界的光合作用启发，科研工作者使用光催化材料诱导化学反应合成有机物[103-106]。光催化材料受到特定的光激发后产生的光生载流子能有效地活化反应物，从而继续发生反应获得高附加值的产物。光催化有机合成与传统热化学合成方法相比有如下优势：可以缩短反应的时间和简化步骤；低能耗、反应更为绿色环保；反应更为温和；不会引入新杂质；等。光催化合成有机物正在逐步地改变着有机化学合成。如图 1-23 所示：在可见光照射下，光催化材料 Pt/TiO_2 成功将硝基苯转化为苯胺并且苯胺选择性可达 95%，同时甘油被氧化成 1,3-二羟基丙酮、甘油醛、甲酸和羟基乙酸等精细化学品[107]。山东大学黄柏标课题组在含有卟啉的 MOFs 结构（PCN-224）中引入 Bi 原子，研究发现：所制备的光催化剂对环氧丙烷与 CO_2 环加成生成的碳酸丙烯酯具有光催化响应[108]，大大降低了反应的能垒，同时 Bi 原子促进了环氧丙烷开环，提高了反应速率（图 1-24）。

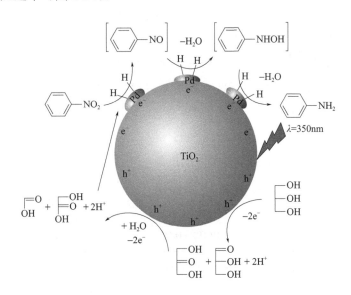

图 1-23　Pt/TiO_2 光催化还原硝基苯和氧化甘油的反应机理示意图

1.2.4.5 光催化还原 CO_2 制备碳氢燃料

根据国际能源署发布的公告，2021 年全球能源燃烧和工业过程中二氧化碳的排放量同比增加 6% 至 363 亿吨，这使大气中 CO_2 等温室气体的浓度大

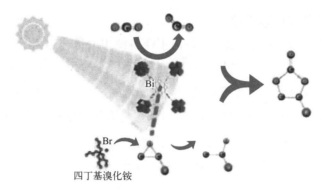

图 1-24　Bi 掺杂的 PCN-224 光催化 CO_2 与环氧丙烷加成的碳酸丙烯酯反应机理示意图

幅上升，温室效应愈加严重，进而危害人类和动植物的生存。所以降低大气中 CO_2 的含量是研究者们普遍关注的问题。受自然界绿色植物光合作用的启发，科研工作者采用光催化技术模拟自然界的光合作用，直接将 CO_2 转化为碳氢燃料[109-110]。Halmann 在 1978 年首次将 p 型磷化镓作为光电极把溶液中的 CO_2 还原为 CH_3OH[111]，从此，光催化还原 CO_2 成为研究热点。光催化还原 CO_2 的反应过程比较复杂，是多电子转移的化学反应过程，产物复杂多样。图 1-25 是可能的反应方程式及相应的标准氧还原电位。理论上，将 CO_2 还原成不同的产物如甲酸（HCOOH）、一氧化碳（CO）等分别需要不同的电子数和不同的氧化还原电位。CO_2 分子在室温下的化学结构稳定性好，因此光催化还原 CO_2 反应的实际应用依旧是一个巨大的挑战。

$$CO_2 + 2H^+ + 2e^- \longrightarrow HCOOH \quad E_{redox} = -0.61 \text{ V vs. NHE}$$
$$CO_2 + 2H^+ + 2e^- \longrightarrow CO + H_2O \quad E_{redox} = -0.53 \text{ V vs. NHE}$$
$$CO_2 + 4H^+ + 4e^- \longrightarrow HCHO + H_2O \quad E_{redox} = -0.48 \text{ V vs. NHE}$$
$$CO_2 + 6H^+ + 6e^- \longrightarrow CH_3OH + H_2O \quad E_{redox} = -0.38 \text{ V vs. NHE}$$
$$CO_2 + 8H^+ + 8e^- \longrightarrow CH_4 + 2H_2O \quad E_{redox} = -0.24 \text{ V vs. NHE}$$

图 1-25　光催化还原 CO_2 生成不同反应物的方程式及对应的标准氧化还原电位

1.3　纳米孔洞金属有机骨架材料的研究背景和发展现状

材料、信息和能源是现代科学技术进步的三大支柱，是人类社会赖以生存和发展的基础。随着社会生产力的发展和科学研究测试手段的提高，各种性能优异的新型材料层出不穷，近期一类新型多孔材料纳米孔洞金属有机骨架（metal organic frameworks，MOFs），也称为多孔配位聚合物（porous coordination polymer，PCP）已成为科学工作者们研究的热点领域之一。如图1-26所示[112]，多孔MOFs材料由无机二级结构单元（金属离子或金属氧化物簇）与有机部分配位构成。这类配位聚合物与传统的无机材料不同，它们具有独特而均匀的孔结构、超高的比表面积和非凡的可加工性。此外它们还是高度结晶的固体材料，有良好的热稳定性和可回收性[113]。多孔MOFs材料的发展可以追溯到配位聚合物的发展[114]。1997年，Buser等研究者通过射线衍射技术确定了普鲁士蓝具体构筑的网络结构[115]。随后科研人员陆续根据普鲁士蓝的结构，合成了一系列金属有机骨架材料，从而受到了广泛的关注[116-119]。1989年，澳大利亚Robson课题组首次将拓扑学应用到配位化合物的合成中，将金属中心称为节点，有机配体定义为连接线，拓扑学的应用为科学工作者们研究多孔MOFs材料带来了很大的方便，这个开创性的研究为多孔MOFs材料的发展翻开了新的一页[120]。1995年，美国密歇根大学课题组选用金属离子与多元羧酸反应，生成了第一个由羧酸配体构筑的配位聚合物，这类多孔配位聚合物第一次被正式命名为多孔MOFs材料，MOFs材料独特的性能使其以难以想象的速度发展着[121]。在过去的几十年里，关于MOFs材料的设计、合成及其应用的文献数量如图1-27所示，呈指数增长[122]，说明了MOFs具有极大的研究价值和潜在的应用前景。

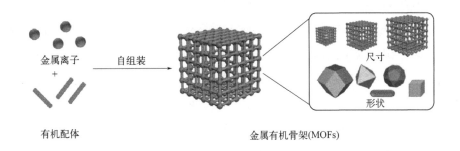

图1-26　MOFs合成和结构示意图

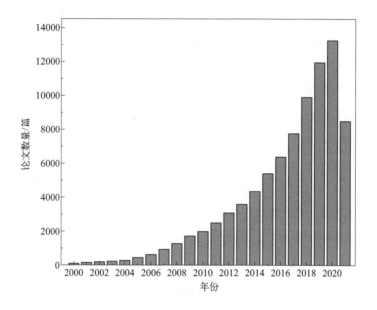

图 1-27　2000~2021 年科学网每年收录的有关
金属有机骨架材料的论文数量

1.3.1　纳米孔洞金属有机骨架材料的特点

纳米孔洞金属有机骨架材料的特点如下。

1.3.1.1　比表面积大

高比表面积的多孔材料在吸附、分离、催化等许多实际应用中具有非常大的优势，所以具有超高比表面积的新型多孔材料是研究的热点，多孔 MOFs 材料因其超高的孔隙率和比表面积而迎来了巨大的发展机遇。MOFs 材料具有永久的孔道结构，孔隙率高达 90% 以上。超高的孔隙率让 MOFs 材料具有超高的比表面积。例如：Koh 课题组同时使用两种有机配体构筑了具有三种孔洞结构的 UMCM-2 材料，比表面积高达 5000$m^2 \cdot g^{-1}$[123]；Hou 课题组成功构筑出比表面积可达 10000$m^2 \cdot g^{-1}$、孔体积可达 4.4$m^3 \cdot g^{-1}$ 的多孔 MOFs 材料[124]，而传统的沸石类分子筛比表面积大约为 10000$m^2 \cdot g^{-1}$，仅为多孔 MOFs 材料的十分之一。非晶的多孔碳材料的比表面积也仅为 3275$m^2 \cdot g^{-1}$，多孔 MOFs 材料具有其他多孔材料无法比拟的高比表面积，因此其在实际应用中发展空间非常广阔。

1.3.1.2 孔径可调

多孔 MOFs 材料不仅具有多孔结构，而且其孔径形状和大小可以根据实际要求进行调控，选择不同长度的、具有不同配位数的有机配体即可制备不同孔径大小的材料。例如 Yaghi 课题组选取了不同的对苯二甲酸衍生物配体和不同长度的有机配体，制备了一系列孔径大小不同但结构相似的命名为 IRMOF-n 的系列多孔 MOFs 材料。这些材料的结构都是永久性的孔洞结构且孔径可调范围为 2.88～3.8 埃（Å，1 Å =10^{-10}m）[125]。所以 MOFs 材料孔径的多样性是传统的多孔材料分子筛所不具有的。

1.3.1.3 结构多样

用于合成 MOFs 的过渡金属离子和有机配体种类繁多，而且，不同的金属离子中心具有不同的氧化态和几何配位要求，形成的次级结构单元也不同，如图 1-28 所示，选用不同的金属节点和配体可以灵活设计 MOFs 的组成和结构。大约每年合成的具有不同结构的 MOFs 骨架高达数千种，在过去的十几年中，由于 MOFs 组分、孔径大小、拓扑结构和功能的灵活性变化，已经有超过 90000 种不同的 MOFs 被研究和报道[126]。目前较为经典的 MOFs 结构包括：IRMOF（isoreticular metal-organic framework）系列、UiO（University of Oslo）系列、HKUST 系列（Hong Kong University of Science and Technology）、MIL（material of Institute Lavoisier）系列、ZIF（zeolitic imidazolate framework）系列、NU（Northwestern University）系列以及 JUC（Jilin University China）系列等。

1.3.1.4 结构可裁剪

多孔 MOFs 材料具有严格的拓扑结构，其结构取决于有机配体的结构和金属团簇的几何形状，所以可以通过改变配体的结构，对 MOFs 的拓扑结构和孔隙率进行合理的设计和预测。如图 1-29（a）所示，利用不同的配体可以得到一系列结构类似而孔径不同的 IRMOFs，孔径大小在几埃到几纳米（最高达 9.8nm）的范围内有规律变化。金属簇 $Cu_2(COO)_4$ 与不同有机配体结合可得到拓扑结构或孔隙结构（曲率和孔隙大小）不同的 MOFs［图 1-29（b），（c）][127]。这种设计原则对开发多功能性的多孔材料具有重要意义，科研人员可以根据不同的性能需求，设计不同的 MOFs 结构。

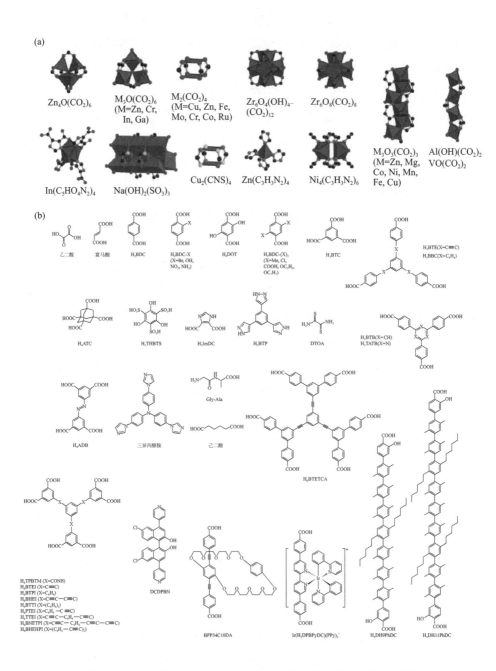

图 1-28 用于构建 MOFs 结构的部分过渡金属离子和有机配体[126]

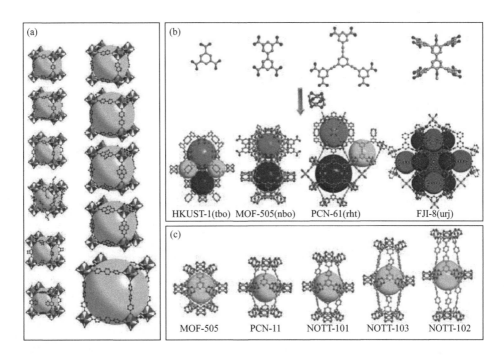

图 1-29　不同有机配体对 MOFs 结构的裁剪

1.3.1.5　具有配位不饱和的金属活性中心

多孔材料在制备过程中，金属离子除了与有机配体发生配位外，还会和溶剂小分子，如 N,N-二甲基甲酰胺（DMF）、H_2O、CH_3OH、CH_3CH_2OH 等发生反应达到配位饱和。当对 MOFs 真空加热处理后，这些溶剂小分子就会从孔洞中脱附出来，从而金属离子变成配位不饱和的状态。MOFs 材料就具有了配位不饱和的金属活性中心，这样的多孔材料比其他类型的催化剂具有更优越的催化性能[128-130]。

MOFs 的结构多样性、超高的孔隙率和比表面积、晶体结构的周期性以及结构的可裁剪性等优点，使其作为一种新兴的多功能材料，在能源储存与转化、气体存储和分离、环境修复、生物医学、催化和传感等方面具有潜在的应用[131-134]。

1.3.2　纳米孔洞金属有机骨架材料的设计与合成

与沸石等其他多孔材料相比，MOFs 材料的特性之一就在于其性能可调性。针对不同应用的需求，MOFs 材料的结构与功能可通过合理选择金属离子或团簇和配体来调控。同时它们连接在一起的方式也对结构与功能有很大影响。研究 MOFs 材料的骨架形成规律可以用来预测合成的 MOFs 材料所具备的性质[135]。深入探讨 MOFs 材料的结构 - 性能关系对合理设计新型功能性的 MOFs 材料具有重要意义。具有不同电子、光学性能和磁性的金属离子/团簇和有机配体可以仔细调整，以匹配特定的应用。

  MOFs 材料大的比表面积有利于反应物的吸附。NU-110 比表面积高达 $7140m^2 \cdot g^{-1}$ [136]。高度有序的孔隙是 MOFs 材料最显著的特征之一。目前所开发的 MOFs 材料大多为微孔（孔径 <2nm）结构，对氢气、二氧化碳等各种气体具有良好的吸附能力。MIL-100（Cr）和 MIL-101（Cr）是非常典型的笼形中孔 MOFs 材料。具体来说，MOFs 材料的孔径可调，范围从微孔到大孔不等，可以容纳不同的功能性物质，例如单个金属原子、纳米粒子、金属配合物、有机染料、多金属氧酸盐等，以改善 MOFs 材料的性能。孔径大小由有机配体的碳链长度或苯环数量决定[137]。科研人员[138]发现：拉长配体结构可以调节 MOF-74 的孔隙大小。不同取代基和官能团可以改变孔的选择性，同时具有独特的化学性质。二价金属如 Zn（Ⅱ）制成的 MOFs 材料具有特殊的孔隙率和广泛的应用前景，但在水或其他恶劣条件下的不稳定性限制了它的实际应用及商业化。MOFs 材料应用于催化领域时，其结构的完整性是达到预期的催化功能的基本条件。研究 MOFs 材料在不同环境中的稳定性，探索其可能的分解途径，开发更稳定的框架结构是现在的研究热点。MOFs 材料在水蒸气或液态水环境中的降解过程可以看作配体被水或氢氧化物取代，因此，阻止这一反应的直接方法是增强金属团簇和配位基团之间配位键的强度。

$$M\text{-}(Ln^-) + H_2O \longrightarrow M\text{-}H_2O + Ln^-$$

  配体 - 金属配位键的强度决定了 MOFs 材料在许多操作环境下的热力学稳定性，根据 MOFs 材料骨架键的强度可以粗略地预测 MOFs 材料的稳定性[138]。配体 - 金属配位键的强度与金属阳离子的电荷正相关，与离子半径负相关。电荷密度高的高价金属离子会形成更强的配位键，形成更稳定的 MOFs 材料，这一规律符合经典的硬/软酸/碱（HSAB）理论[139]。根据 HSAB 理论，硬路易斯酸和碱、软路易斯酸和碱之间的相互作用要比硬酸和软碱、软酸和硬碱之间的相互作用强得多。为了获得稳定的 MOFs 材料，优先选择羧酸基配体（硬路易斯碱）和高价金属离子（硬路易斯酸），或唑基配体（软路易斯碱）和低价过渡金属离子（软路易斯酸）构建框架。例如，羧酸基配体是硬碱，与高价金属离子[Ti(Ⅳ)、Zr(Ⅳ)、Al(Ⅲ)、Fe(Ⅲ)、Cr(Ⅲ)]共同形成较稳定的 MOFs 材料，如基于 Al(Ⅲ)、Fe(Ⅲ)和 Cr(Ⅲ)的 MIL 系列金属有机框架，包括著名的 MIL-53、MIL-100 和 MIL-101。简单的合成途径、在水甚至酸性条件下的强稳定性和低毒性的特点最终推动了这系列材料的实际应用。在此策略的指导下，获得了数十个具有良好稳定性的金属有机框架（图 1-30）。

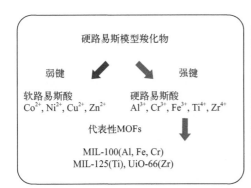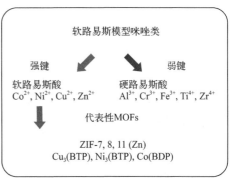

图 1-30　基于 HSAB 理论构建稳定金属有机框架的策略

随着不同领域的科研工作者对 MOFs 材料研究的逐渐深入，MOFs 的合成方法也逐渐多样化。根据文献报道，目前的合成方法主要有：水（溶剂）热法、机械研磨法、溶剂扩散法、离子液体法、超声与微波法以及后合成法等。采用不同的合成方法对材料的性能影响还是挺大的，所以根据不同的要求采用合适的制备方法。

1.3.2.1　水（溶剂）热法

水（溶剂）热合成法是目前制备 MOFs 材料最为常用的方法[140]，具体方法：将所有的前驱体包括金属离子和有机配体溶于相应的溶剂中，然后将所制备的反应物全部转移到反应釜中，在高温和高压条件下，金属盐和有机配体可完全溶解，短时间内快速成核生成金属有机框架材料。反应釜内的压力与反应釜的容积和溶剂所占的体积直接相关，所以，为控制压力，有机溶剂所占体积应小于反应釜总容量的 75%，水溶剂所占体积应小于反应釜总体积的 80%。反应温度一般控制在 300℃以内，根据不同的合成要求选择不同的溶剂，通常选用高介电常数和极性的氢离子供体，例如水、甲醇、乙醇和醋酸；或非氢离子供体，如 N,N- 二甲基甲酰胺、乙腈和丙酮。这种方法优点是：常温常压下难溶的前驱体溶解度增大，反应活性增加，从而促进了反应的进行；选用不同的溶剂，反应物的物理化学性质会有所改变，最终产物的结构也有所不同，所以可以通过选择不同的反应溶剂改变产物的结构，最终获得目标产物。这种方法缺点是：观测不到具体的反应历程，研究机理非常困难。但是合成出来的 MOFs 晶体结构完整，所需设备简单，所以这种方法得到研究者们的广泛采用。

1.3.2.2 溶剂扩散法

溶剂扩散方法是在常温常压下进行的，反应条件比较温和。常用的扩散方法，将前驱体分别溶解在密度不同的溶剂中，然后分别放入反应容器中，静置，这两种溶液中的前驱体慢慢扩散，并在两种溶液的界面处反应长出晶体。另外一种情况是：在两种不同的反应溶液之中加入第三种易挥发的溶剂，第三种易挥发的有机溶剂扩散到前驱体的混合液中，促使反应进行得到产物[141]。这种方法的优点是：合成材料的条件温和而且可以获得高质量单晶材料。缺点是：需要反应的前驱体在室温下就不容易溶解并且反应需要的时间非常长，甚至需要几个月的时间。

1.3.2.3 机械研磨法

上述方法由于成本和时间的关系都不适用于大规模的工业生产要求，为了让 MOFs 真正地走出实验室，科研工作者首先就想到了可以大规模生产的机械研磨法。机械研磨法：将合成的前驱体固体金属盐和固体有机配体放在球磨机里进行研磨，在研磨过程中，前驱体之间的反应依赖于对机械能的直接吸收。球磨过程中，化学反应所需的能量来自于球和反应物之间较高的摩擦和冲击。高能研磨引起结构应力的改变、键断裂和形成活性自由基，反应原子层暴露，引起固体反应物界面的化学反应[142]。Pichon 研究组第一次用机械研磨的方法成功地合成了铜与异烟酸配位生成的 MOFs 材料[143]。现在已经有不少用机械研磨法制备的 MOFs 材料了[144]。机械研磨法优点是合成需要的时间短，适合大规模生产；缺点是合成的过程中可能含有的杂质较多，目前还不成熟，不能广泛用于制备 MOFs 材料。

1.3.2.4 离子液体法

离子液体法是制备 MOFs 材料的一种新型的有效方法。离子液体是指在室温下完全由自由移动的阴阳离子组成的液态低温熔融盐[145]，具有不可燃、不挥发、导电能力好、热容大、稳定性强等特性。离子液体对无机盐和有机配体有很好的溶解性能，离子液体法未来会在MOFs材料的合成中展现出重要作用。

1.3.2.5 超声化学法

超声化学合成方法是利用超声波（20～1000kHz）的能量加速金属离子与有机配体之间的反应[146]。该方法优点是：高效、低成本且环保。超声波作用于前驱体溶液时，反应液中产生气泡并振荡，能量在这些气泡中累积，气泡尺寸增长到一定的范围后在短时间内破裂，产生巨大的能量，能量通过气泡破裂分布到周围环境中，促使局部温度（高达500K）和压力（高达100MPa）急速上升，强化和加速成核反应。超声的时间、功率和温度对MOFs材料的性能有一定的影响。另外，MOFs晶体的形貌、尺寸和产率主要取决于前驱体的初始浓度和反应时间。晶体的大小直接依赖于前驱体的初始浓度，当前驱体初始浓度较低时，虽然MOFs材料的产率随着反应时间的进一步增加而增加，但是晶体尺寸会较大且形状不规则，反应时间的增加使得MOFs晶体尺寸和产率增加。与传统水热法相比，超声法使成核中心均匀，结晶时间缩短，颗粒尺寸明显更小。在水（溶剂）热合成法中，超声波可作为辅助工具。

此外，其他的合成的方法，如微波法、溶胶凝胶法、电加热法，等等，就不一一介绍了。

1.3.3 纳米孔洞金属有机骨架在光催化领域的应用及优化

MOFs这类由无机二级结构单元（金属氧化物簇或金属离子）与有机配体配位构成的多孔材料，由于其独特的性能受到广泛的关注。其中MOFs材料模块化和容易在其孔道中负载光活性催化位点的特性，使MOFs基光催化剂成为最近的研究热点[147]。MOFs材料这种高度结晶的固体材料同时具有多相和均相催化剂的一些催化优点（良好的热稳定性，可容易地从气相和液相中分离出来继续循环使用），因此MOFs基复合材料在光催化领域展现出巨大的潜力[148-149]。MOFs材料中定义明确的有机分子组件可以被认为是排列在晶格中的有机分子催化剂[150]，可以将具有高催化活性和选择性的有机金属配体引入MOFs的框架中。MOFs具有高比表面积，可以将大量催化活性位点引入

MOFs 的框架中，提供与均相催化剂相当的活性和选择性[151]。MOFs 材料的配位不饱和金属还可作为氧化还原反应的单点催化反应中心，或作为锚定催化活性位点的结合位点。MOFs 材料有机配体、金属簇或空腔内的受限催化中心也可产生单中心催化反应位点[152-154]。MOFs 材料已经显示出了作为连接均相催化剂与多相催化之间桥梁的巨大潜力，为探索各种具有稳定性和高光催化活性的新型光催化剂打开了大门（图 1-31）。

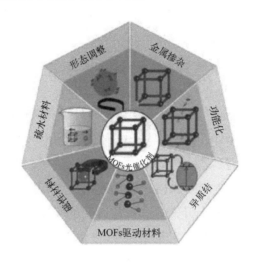

图 1-31　MOFs 基光催化剂

1.3.3.1　基本原理

在光催化过程中，不同于传统的半导体光催化剂，MOFs 材料有独特的配体 - 金属电荷转移（LMCT）机制[155]。在光照射下，MOFs 材料中的有机配体可以作为光敏剂吸收光能并被激发产生电荷。光生电荷从 MOFs 材料的最高占据分子轨道转移至最低未占据分子轨道，最后转移到金属氧簇的表面，而不是整个 MOFs 材料的表层，这提高了光生载流子的分离效率。例如在 MIL-125（Ti）中，2-氨基对苯二甲酸基团被光激发产生了光电子，随后通过还原 Ti（Ⅳ），将光生电子转移到钛氧簇上形成 Ti（Ⅲ）[156]。此外，金属 - 配体电荷转移、配体 - 配体电荷转移和金属 - 金属 - 配体电荷转移机制也被用来解释各种光催化过程[157-158]。对于铁基 MOFs 材料来说，Fe-O 簇可直接被光激发产生光生载流子[159]。氧气与光生电子产生超氧自由基，光生空穴直接氧化有机污染物或者与水分子反应生成羟基自由基，而且 MOFs 材料比表面积大且孔道结构排列有序有利于活性分子的引入，更有利于电荷分离。

1.3.3.2 金属中心和有机配体的优化

MOFs 由金属离子/簇和有机配体组成，可通过灵活调整 MOFs 的结构来增强光吸收范围[160-161]，MOFs 的金属部分几乎涵盖了所有的金属。其中类似 Ti^{4+}/Ti^{3+}、Zr^{4+}/Zr^{3+} 和 Fe^{3+}/Fe^{2+} 的可变价态的金属离子与羧酸配体的氧原子配位可以形成金属-氧簇化合物。这些金属离子能够吸收光激发电荷，由基态转变为激发态，参与光催化反应，因此，变价金属离子基 MOFs 材料具有一定的还原能力，是很好的光催化材料[162-166]。如图 1-32 所示，在可见光的照射下，NH_2-MIL-125（Ti）中的氨基功能化的配体吸收可见光产生了光电子，随后转移到团簇上，将 Ti^{4+} 转变成 Ti^{3+} 活性位点，NH_2-MIL-125（Ti）吸附的 CO_2 被 Ti^{3+} 活性位点还原成甲酸，三乙醇胺（TEOA）提供电子使 NH_2-MIL-125（Ti）回复到基态[167]。这个实验表明：MOFs 材料中的变价金属中心是实际的催化活性物种。铁基 MOFs［例如 MIL-101（Fe）、MIL-53（Fe）和 MIL-88B（Fe）］也具有良好的催化活性[168]。这是因为 Fe-MOFs 中的配位不饱和 Fe 位点可作为氧化还原反应位点直接参与反应（图 1-33）。研究者利用 Cu^{2+} 和三咪唑硼

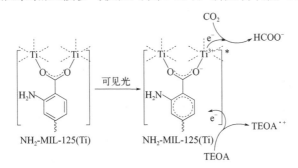

图 1-32　在光照条件下 NH_2-MIL-125（Ti）光催化原理图

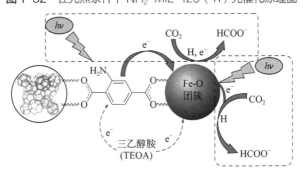

图 1-33　Fe 基 MOFs 光催化原理示意图

氢 BH(im)$_3^{-}$ 合成了 BIF-29 纳米笼（图 1-34），BIF-29 具有优异的光催化活性[169]。BIF-29 中大量的不饱和 Cu 位点提高了材料本身的电荷分离效率，稳定 COOH 中间体，所以 BIF-29 具有高的催化活性和选择性。

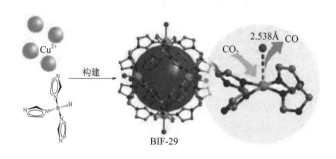

图 1-34　BIF-29 的合成和光催化原理示意图

有机配体对配体-金属电荷转移有作用，从而影响 MOFs 材料的光响应能力、稳定性和催化位点[170-171]。改性有机配体是制备新型高效 MOFs 材料的有效方法。以对苯二甲酸（H_2BDC）有机配体为例，以对苯二甲酸为原料合成的 MIL-125（Ti）仅在紫外光下响应[172]，以改性后的对苯二甲酸（接上一个—NH_2 基团）为原料，用相同的方法合成了同结构的 NH_2-MIL-125（Ti），对可见光的吸收区域增加到了 520nm[173]。研究者[174] 从实验和理论两方面对 NH_2-MIL-125（Ti）进行了详细的研究：最高占据分子轨道能级的改变是光学特性增强的主要原因。引入单个的—NH_2 基团使最高占据分子轨道的能级升高 1.20eV，最低未占据分子轨道没有变化。另外，接入其他官能团包括—OH、—CH_3、—Cl 以及 BDC-(NH_2)$_2$ 的 MIL-125（Ti）带隙宽度也会变化，但是效果不明显。如果采用具有较高 π 共轭基团的类染料修饰的对苯二甲酸作为原料制备的 MR-MIL-125（Ti），则其吸收光谱有明显的光吸收红移，带隙宽度为 1.93eV[175]。如图 1-35 所示，将 Re(CO$_3$)(bpy)Cl（bpy=2,2-联吡啶）光敏分子引入 MOFs 中制备 Re 掺杂的多相光催化剂 UiO-67，大大提高了其光催化活性[176]。Kajiwara 课题组将 Ru-CO 配合物掺入 UiO-67 中制备 UiO-67-Ru。结果表明：UiO-67-Ru 的催化活性优于光敏配合物[177]（图 1-36）。除了合成新的有机配体，混合配体构筑更高效的 MOFs 光催化剂同样引起了研究人员的关注，这是因为 MOFs 结构的多功能性和灵活性[178-179]。例如，在原 MOFs 材料中通过溶剂热法或后合成修饰方法加入其他的有机配体制备新型 MOFs 材料[180]。混合组分的 MOFs 材料活性位点种类多，光催化性能得到提升。Lee 课题组用 2, 3- 二羟基对苯二甲酸（CATBDC）取代 UiO-66 中的部分 BDC 配体，制备 MOFs 光催化剂（UiO-66-CrCAT 和 UiO-66-GaCAT）[181]。UiO-66-CrCAT 和 UiO-66-GaCAT 都具有高催化活性（图 1-37）。

图 1-35　UiO-67 的结构模型和光催化原理示意图

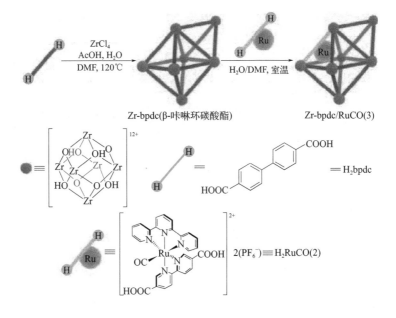

图 1-36　UiO-67-Ru 的合成示意图

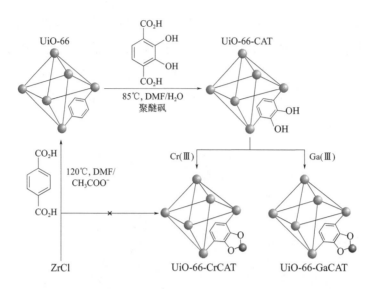

图 1-37　UiO-66-CrCAT 和 UiO-66-GaCAT 的合成路线图

1.3.3.3　构建异质结

抑制光生载流子的复合，提高光生电荷的分离和转移效率是提高催化材料催化活性的重要方法。将 MOFs 与其他半导体材料耦合形成异质结结构或电子俘获位点，促使光生电荷的转移，有效实现电荷的分离，是有效提高 MOFs 材料光催化活性的策略之一。MOFs 的多孔网络可以作为载体促进半导体的分散，从而暴露更多的活性位点。另外，异质结可以有效地促进电荷分离。根据价带、导带的位置以及半导体的 n、p 类型（图 1-38），异质结型光催化材料构型一般可分为 I 型、II 型、Z 型和肖特基异质结。其中，II 型异质结在光催化过程中，半导体 A 导带上的光生电子会转移至半导体 B 的导带上，半导体 B 价带上的光生空穴转移至半导体 A 的价带上，促使光生电子和空穴有效分离。Z 型异质结在光催化过程中，半导体 B 导带中的光生电子转移至半导体 A 的价带中与光生空穴结合，实现光致电荷分离。肖特基异质结将助催化剂链接到半导体上，半导体作为电子陷阱，有效地捕获光生电子，提高光生电荷的分离效率，从而提高光催化活性。如图 1-39，Wang 等科研人员将半导体 $g-C_3N_4$（或 CdS）与 Co-ZIF-9 复合制备复合光催化材料[182-183]。实验结果表明：复合催化剂显示出比其组成半导体更好的光催化活性，半导体和 MOFs 材料之间的直接接触促进了光电子的转移。Liu 等人在 Zn_2GeO_4 纳米棒表面生长 MOFs 材料 ZIF-8，制备半导体 MOFs 复合光催化材料。结果表明：复合材料表面包裹了一层 MOFs 材料，增加了其吸附反应物的能力；同时扩大了复合材料的光吸收范围，从而提高了光催化活性[184]。Crake 等人采用原位生长法合成 TiO_2/NH_2-UiO-66 复合材料，与纯二氧化钛相比，复合材料的光催化活性显著增强（图 1-40），这是因为 NH_2-UiO-66 和二氧化钛之间形成的 Ⅱ 型异质结有效促进了界面光生电荷的转移[185]。如图 1-41，将 $CH_3NH_3PbI_3$（$MAPbI_3$）钙钛矿量子点引入到铁卟啉基 MOFs PCN-221（Fe_x）的孔道中制备复合光催化材料，结果表明：复合材料有优异的稳定性；同时光催化活性与 PCN-221 相比，提高了 38 倍[186]。

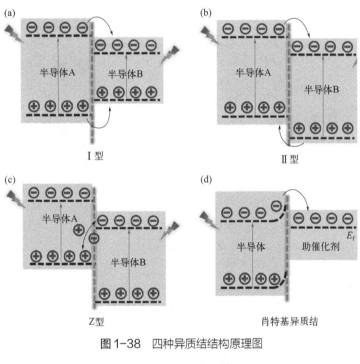

图 1-38 四种异质结结构原理图

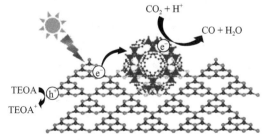

图 1-39 Co-ZIF-9 和 g-C_3N_4 的光催化原理示意图

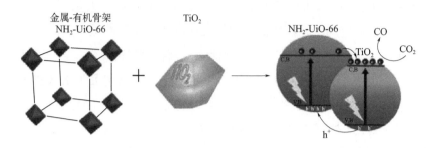

图1-40　TiO$_2$/NH$_2$-UiO-66复合材料的制备流程图和光催化原理示意图

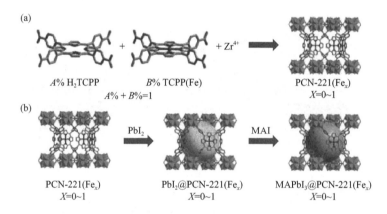

图1-41　PCN-221（Fe$_x$）的制备流程图（a）和MAPbI$_3$@PCN-221（Fe$_x$）的合成示意图（b）

1.3.3.4　金属负载

将金属纳米颗粒负载到MOFs材料上，构成复合光催化剂的优势：利用金、银、铜等金属局域表面等离子体共振拓宽符合光催化材料的光吸收范围，以及等离子体中电子从金属表面转移至MOFs材料框架中，促进光生载流子分离，从而提高光催化剂的活性[187-188]；利用金属纳米颗粒与MOFs材料之间的电势差在接触界面产生内建电场，内建电场促使光生电荷转移至金属表面，提高光生电荷的分离与转移效率[189]。科研工作者采用原位法合成了一系列含Co量不同的Co-ZIF-9/TiO$_2$复合光催化材料[190]。结果表明：二氧化钛和Co-ZIF-9两者之间紧密结合，促进光生电子转移，电荷分离效率更高，催化活性更好（图1-42）。

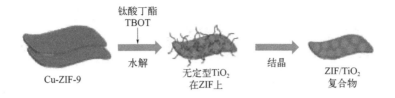

图 1-42 Co-ZIF-9/TiO$_2$ 复合材料的合成示意图

1.4 铋系半导体光催化材料的研究背景和现状

目前，光催化技术是解决能源问题和缓解环境问题的有效方法。过去几十年，科研人员努力开发新材料、新技术使其催化效果达到最大化，以实现其实际应用，例如，处理有机废水、光催化产氢、染料敏化太阳能电池等[191-195]。但是光催化技术的实际应用，关键技术还是在高效的光催化材料选择与制备上。应用广泛的 TiO$_2$ 光催化剂带隙较宽（3.2eV），仅能吸收太阳光中的紫外光，使其实际应用受到极大限制[196-199]。可见光在太阳能中占比较高（45%），所以探索制备高效的可见光光催化剂更有实际意义。

被称为"绿色元素"的铋（Bi）是有低毒性和低放射性的重金属元素，全球储量丰富而且中国是铋资源储量世界第一的国家，占世界总储量的 70% 以上。铋及其化合物在医药、航天、电子、冶金、化工等领域发挥重要的作用，因此开发和高效利用铋系半导体材料具有重要的学术价值和实际意义。如图 1-43 所示，铋系光催化剂正成为研究的热点，其相关文献数量这十几年激增。

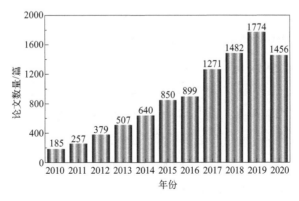

图 1-43 2010 年至 2020 年发表的有关铋系光催化剂的论文数量

Bi 独特的 $6s^2$ 电子构型使其多数化合物带宽较窄,是一类对可见光响应的光催化剂。Bi 与 O 形成氧化物时,Bi 的 6s 轨道与 O 的 2p 轨道共同杂化,如图 1-44 所示,提升了价带位置从而缩小带隙,同时降低了轨道对称性,产生了偶极子、压电、铁电、非线性光学等性能,大多数铋基半导体的价带位置较正。而且生成的铋基氧化物大多数为层状材料,片层间的内建电场有利于光生载流子的分离,因此铋基氧化物是非常理想的光催化材料[200]。

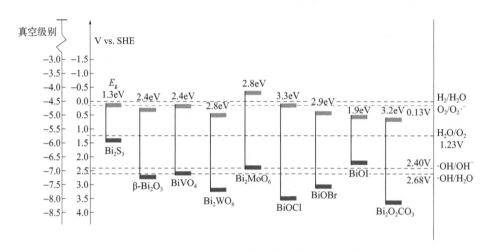

图 1-44　常见 Bi 基光催化剂的能带结构示意图

1.4.1　铋系光催化材料

1.4.1.1　卤氧化铋（BiOX）

间接带隙卤氧化铋（BiOX,X=F、Cl、Br、I）属于三元组分（Ⅴ-Ⅵ-Ⅶ）半导体,具有较好的化学稳定性和光学性能,四方晶系,$P4/nmm$ 空间群[201-202]。如图 1-45 所示,BiOX 是由 [X-Bi-O-Bi-X] 结构堆积而成,是 $[Bi_2O_2]^{2+}$ 层和与双 $[X]^-$ 层排列的二维层状结构[203]。层间原子的范德华力使其产生高度各向异性的光学特性。卤氧化铋（图 1-46）的这种独特结构使原子和轨道得到充分的极化,促使带正电的 $[Bi_2O_2]^{2+}$ 层和带负电的卤素原子层形成一个内部静电场,这个内建电场能提高层间电荷迁移能力,促使光生载流子在层间分离,从而提升材料的光催化性能[204]。如图 1-47 所示,随着卤素原子量的变大,BiOX 的禁带宽度逐渐变小,光吸收范围从紫外光扩展至近红外光区[205]。带隙合适的 BiOI 和 BiOBr 吸收可见光和近红外光，逐渐成为研究的热点。例如，如图 1-48 所示，Chen 等在室温下，制备了花状 BiOX 分级结构，这种三维的分级形貌有利于光的吸收，提高了 BiOX 的可见光降解活性[206]。Huang 课题组将 BiOI 与 g-C_3N_4 复合构建成 p-n 型异质结复合材料，加入 BiOI 提高了对可见光的吸收，有效促进了光生载流子的分离，提高了 g-C_3N_4 的可见光催化性能[207]。Huo 等人通过调控溶剂比例制备分级花状 BiOBr 微球，这种形貌使其光催化性能显著增强[208]。

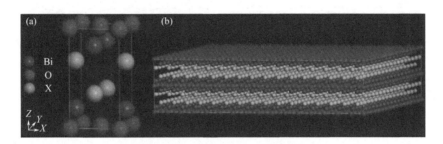

图 1-45　BiOX 的晶体结构模型
（a）单胞；（b）二维层状

图 1-46　制备的卤氧化铋的形貌图片

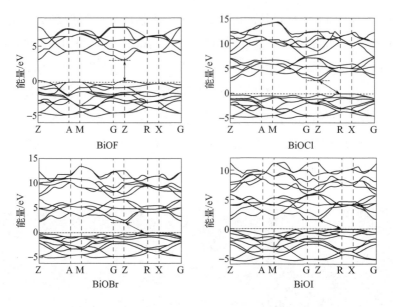

图1-47 BiOX（X=F、Cl、Br、I）的能带结构

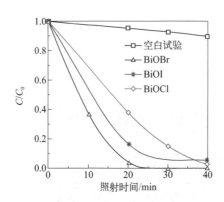

图1-48 3D花状BiOX可见光降解图谱

1.4.1.2 氧化铋（Bi_2O_3）

氧化铋（Bi_2O_3）结构简单，带隙为$2.1 \sim 2.8eV$，是可见对光响应的半导体光催化材料。如图1-49所示，Bi_2O_3有6种晶相：单斜相α相；四方相β相；体心立方相γ相；面心立方相δ相；三斜相ω相和ε相。其中ω相和ε相不稳定，六种晶相在一定的条件下可互相转化[209]。研究表明：半导体的晶相会影响对应的光催化活性，其中β-Bi_2O_3在六种晶相中具有最好的光催化活性。例如，如图1-50所示，Xiao等研究者制备的β-Bi_2O_3纳米球在可见光下具有

极佳的光催化性能，优于 P25-TiO$_2$ 和市售的 Bi$_2$O$_3$[210]。此外，一种新型铋系氧化物 m-Bi$_2$O$_4$ 被报道，其光吸收波长可到 620nm，光催化活性和稳定性要比 Bi$_2$O$_3$ 更好[211]。Bi$_2$O$_3$ 作光催化剂单独使用存在两大缺陷：一是光生载流子分离效率低，光催化性能差；二是在反应过程中会产生严重的光腐蚀从而转变成次碳酸铋（Bi$_2$O$_2$CO$_3$）[212]。因此其常与其他半导体复合形成异质结复合光催化材料。

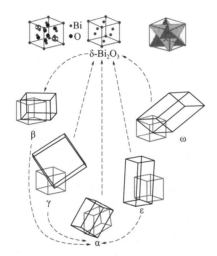

图 1-49　Bi$_2$O$_3$ 不同晶型之间的转换

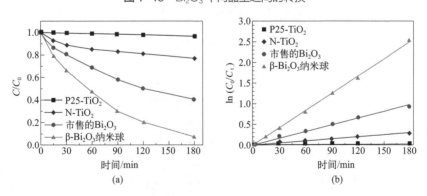

图 1-50　不同催化剂光降解对乙酰氨基酚（APAP）的性能研究（a）和 ln(C_0/C_t）与光照时间的线性拟合图（b）

1.4.1.3　钨酸铋（Bi$_2$WO$_6$）和钼酸铋（Bi$_2$MoO$_6$）

组成最简单的奥里维里斯（Aurivillius）型铋系氧化物半导体 Bi$_2$XO$_6$（X=Mo、W）的晶体结构属于正交晶系，类似于三明治层状结构，由 [Bi$_2$O$_2$]$^{2+}$

层和钙钛矿 XO_4^{2-} 层交替排列组成的一类窄带隙半导体[213-214],如图 1-51 所示。该类材料价带主要是由 Bi 的 6s 和 O 的 2p 轨道杂化形成的,导带以 W(或 Mo)的 5d(或 4d)轨道为主[215]。Bi_2WO_6 和 Bi_2MoO_6 的光学带隙约为 2.7eV,最大吸收波长可达 500nm 左右,是良好的对可见光响应光催化材料[216]。1999 年 Kudo 和 Hijii 首次通过固相合成工艺得到了 Bi_2WO_6,并将其用于光解水产氧的实验[217]。随后 Ye 等人制备 Bi_2WO_6 用于降解有机污染物,实验结果表明,Bi_2WO_6 是不仅可有效光解水还可降解有机污染物的可见光催化材料[218]。简易的高温固态烧结合成工艺制备的 Bi_2WO_6 颗粒尺寸大且不均匀,光催化活性低。后来研究者通过改变制备方法,如水热/溶剂热法、溶胶-凝胶法等来合成具有小尺寸特殊形貌的 Bi_2WO_6,Bi_2WO_6 的光催化性能得到有效的提高[219]。如图 1-52 所示为水热工艺制备 Bi_2WO_6 微纳米结构示意图[220]。

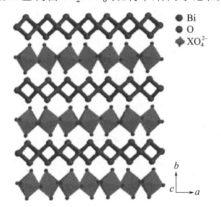

图 1-51 Bi_2XO_6(X=Mo、W)的晶体结构示意图

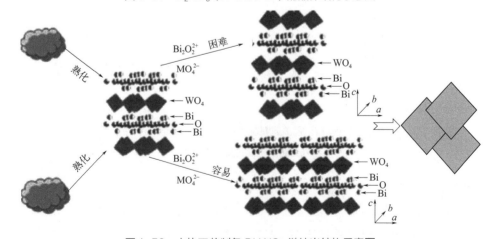

图 1-52 水热工艺制备 Bi_2WO_6 微纳米结构示意图

Wang 等人在强酸性（pH=1）条件下采用水热反应合成了超结构花状 Bi_2WO_6[221]，光催化性能明显优于固相合成得到的 Bi_2WO_6（图 1-53）。此外，好多课题组采用掺杂改性、复合修饰等方法提高 Bi_2WO_6 的光催化活性，也取得了很好的效果。如图 1-54 所示，Zhang 等研究者制备了新型环化聚丙烯腈 c-PAN 修饰 Bi_2WO_6 的复合光催化材料。研究结果表明：c-PAN 和 Bi_2WO_6 结合形成的不饱和 N 位可快速捕获和激活 N_2 分子的反应活性位点，因而复合

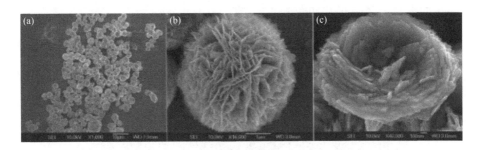

图 1-53 超结构花状 Bi_2WO_6 的 SEM 图

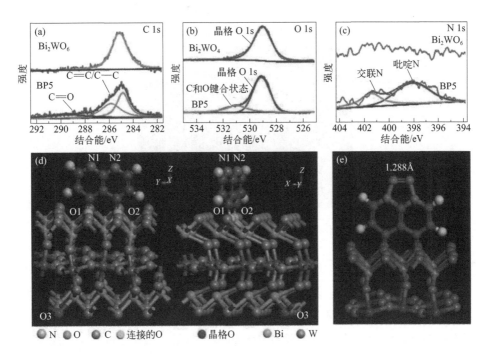

图 1-54 c-PAN/Bi_2WO_6 和 Bi_2WO_6 的 XPS 能谱 C 1s（a）、O 1s（b）、N 1s（c）及 c-PAN/Bi_2WO_6 的结构优化图（d）和吸附 N_2 的结构图（e）

材料比纯组分的 Bi_2WO_6 具有更优异的光催化固氮性能[222]。Kudo 等最早制备了 Bi_2MoO_6，并将其作为可见光催化材料催化分解水制氢，并研究其光催化活性的影响因素：能带结构和结晶度[223]。自此以后，Bi_2MoO_6 逐渐受到关注（图 1-55、图 1-56）。Xia 等人利用水热法制备了新型碳量子点 CQDs/Bi_2MoO_6 杂化材料，光催化性能比纯 Bi_2MoO_6 要好，对罗丹明 B、四环素盐酸盐（TC）等有良好的降解活性，这归因于作为电子受体的 CQDs 提高了界面电荷转移能力，从而提升了材料的电荷分离效率[224]。

图 1-55　水热法制备 Bi_2MoO_6/ 碳量子点的形貌图

图 1-56　水热法制备 Bi_2MoO_6 的形貌图

1.4.1.4　钒酸铋（$BiVO_4$）

$BiVO_4$ 是一种无毒、稳定的三元氧化物半导体光催化材料，制备方法简单廉价，可以进行大规模合成[225]。$BiVO_4$ 作为钒铋矿在自然界中以正交晶系结构存在，在实验室合成的主要有三种晶型：单斜白钨矿（m）、四方硅酸锆型（tz）和四方白钨矿（ts），如图 1-57 所示[226-228]。研究表明单斜相 $BiVO_4$ 是直接带隙半导体而且带隙适中（约 2.4eV），对可见光响应性最好，因此在可见光催化方面非常具有应用前景（图 1-58）[229]。如图 1-59 所示，研究者采用表面活性剂为形貌控制剂水，热法制备了 {010} 晶面优先取向的 m BiVO$_4$ 片状纳米光催化材料，与 m BiVO$_4$ 纳米棒和微晶相比，暴露 {010} 晶面的 m BiVO$_4$ 纳米片可见光催化降解 RhB 和分解水产 O$_2$ 的活性大大增强[230]。另外，Sun 等人在油酸钠条件下采用水热法合成直径为 5nm 和管壁厚 1nm 的介孔结构 m BiVO$_4$ 量子管，其具有显著的光和温度的双响应性质，可以应用于温度传感器和可见光催化领域（图 1-60）[231]。

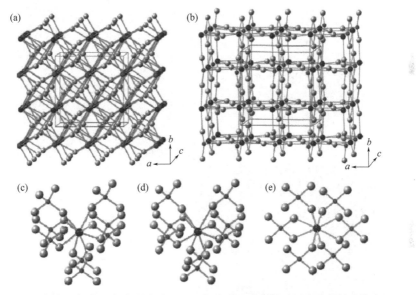

图 1-57　四方白钨矿（a）四方硅酸锆型 BiVO$_4$（b）的晶体结构（深色直径较小的表示 V，深色直径较大的表示 Bi，灰色表示 O）及四方白钨矿（c）单斜白钨矿（d）四方硅酸锆型 BiVO$_4$ 结构中 V 和 Bi 离子（e）的局部配位

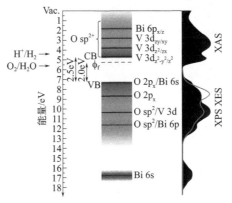

图 1-58　BiVO$_4$ 的能带结构示意图

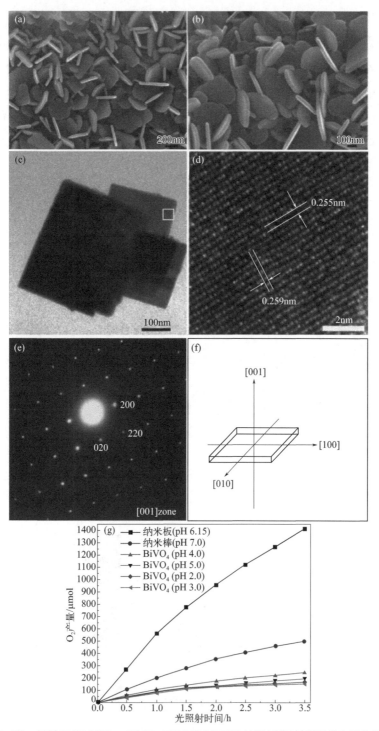

图 1-59　暴露 {010} 晶面 m BiVO$_4$ SEM 和 HRTEM 图片以及其增强的光催化性能

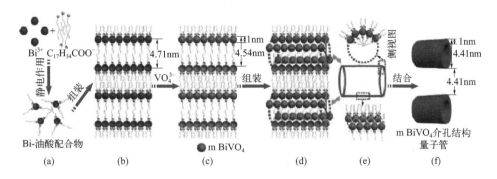

图 1-60　Sun 等报道的孔结构 m BiVO₄ 量子管的合成机制

1.4.1.5　其他铋系半导体

硫化铋（Bi_2S_3）是一种禁带宽度大约 1.3eV，全光谱响应型的层状光催化半导体材料，但是单一 Bi_2S_3 的光催化性能较差，这主要是其价带位置为负，导致光生空穴的氧化能力较弱，光生载流子复合率高。科研工作者把目光主要放在了 Bi_2S_3 复合材料的制备上[232]。其他含铋光催化剂还有钛酸铋（$Bi_4Ti_3O_{12}$）、磷酸铋（$BiPO_4$）、甲酸氧铋（$BiOCOOH$）等，光催化活性普遍较弱，研究不多。

1.4.2　铋系半导体复合光催化材料

铋系半导体材料作为光催化剂在很多应用领域被广泛研究，但是限制其应用的主要原因是催化活性不够高。如图 1-61（a）所示，在光催化反应过程中，光生载流子的分离和迁移过程直接决定光催化材料的催化能力。很多单组分铋系半导体材料的多数光生载流子在迁移的过程中复合，很少一部分成功参与催化反应，从而降低了其光催化性能。大量研究表明制备复合光催化材料是提升光催化性能和拓宽光吸收范围的有效办法之一[233-234]。就铋系半导体而言，如图 1-61（b）所示，与另一组分材料（如贵金属、氧化物半导体等）复合构建异质结有效促进光生载流子的分离和迁移，是提升其光催化活性的有效途径。

1.4.2.1　铋系半导体 / 贵金属复合光催化剂

铋系半导体与贵金属复合是提升铋系半导体材料光催化活性的一种有效手段。一方面贵金属的功函数相对较低，费米能级位置也较低。贵金属与铋系半导体复合后，两者界面处产生肖特基势垒。当吸收光时，半导体的光生电子

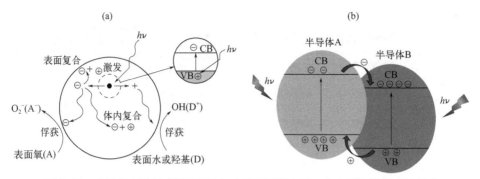

图1-61 半导体光催化过程示意图（a）和异质结电子-空穴对的分离过程（b）

向贵金属中迁移，贵金属像"电子陷阱"，这样就提高了半导体中的光生载流子的分离效率，从而提升了复合材料的光催化活性[235-238]。另一方面利用贵金属的表面等离子体共振效应（SPR），将贵金属负载到半导体颗粒或阵列结构的表面，或以贵金属为核心构建金属/半导体核壳结构，或将半导体纳米晶串联到一维贵金属表面形成项链状金属/半导体复合结构，可以有效地提高半导体光催化材料的可见光利用效率，并促进光生载流子的高效分离。贵金属纳米粒子在复合材料中的作用会因半导体带隙及入射光波长的不同而有所区别（见图1-62）[239-240]，贵金属的表面等离子体共振效应可以吸收太阳光谱中的可见光，从而给半导体材料提供光电转换或分解水制氢所需的电子。Zhang等研究者在$BiVO_4$纳米纤维表面合成了碳纳米层和纳米Ag[241]。如图1-63所示，由于Ag的"电子陷阱"作用，$BiVO_4@C/Ag$复合光催化材料的催化活性优于$BiVO_4@C$。Niu等人在$BiFeO_3$颗粒表面生长了纳米Pt，制备了$Pt/BiFeO_3$复合光催化剂。实验结果表明：复合光催化材料提高了光生载流子的分离效率，光催化活性随之提高（图1-64）[242]。

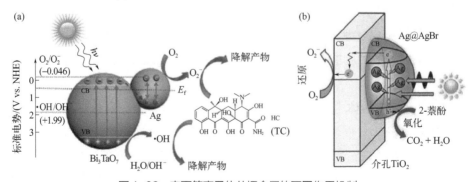

图1-62 表面等离子体共振金属的不同作用机制

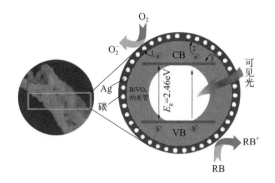

图 1-63　BiVO$_4$@C/Ag 复合光催化剂降解罗丹明 B 的机理图

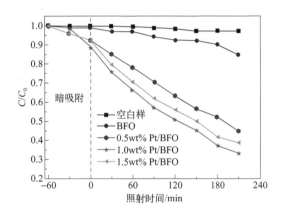

图 1-64　Pt/BiFeO$_3$ 复合光催化降解甲基橙的效率图

最近，单质金属铋（Bi）以其独特的物理化学性质及优良的光催化性能而引起人们极大的关注[243]。铋具有半金属性质，在其导带和价带之间有少部分的能量重叠（约 38meV），因而在无光时其导带中就有一定浓度的电子，价带中也有等量的空穴。尽管金属铋很早就已经被应用于超导材料、压电陶瓷、医学领域及润滑油添加剂等中，但是将单质铋应用于半导体光催化领域是近几年才开始研究的方向，而且以前的报道大多是将其负载于光催化剂粉体表面用于富集光生电子。Jiao 等人发现金属铋具有很强的表面等离子体共振效应，其在 270nm 和 390nm 有两个较强的表面等离子体共振吸收峰，而吸收带边波长可以一直延伸至 650nm，且其能带结构和光吸收特性可以通过改变纳米颗粒的尺寸及维度进行调控，因而金属铋有作为表面等离子体共振金属代替贵金属材料的物理属性。

将金属铋与半导体材料,尤其是一维半导体阵列复合,可以有效地拓宽半导体对太阳光的响应范围并提高光生载流子的分离效率,从而提高半导体的光催化及光电转换性能。一维半导体阵列结构相比于纳米颗粒具有更高的光子捕获效率,更快的电子传输路径以及更有利于电子的定向传导。在光照条件下一维半导体阵列内部由于电子-空穴的定向流动而存在较强的内建电场,内建电场可以诱导相邻纳米棒之间的金属颗粒相向生长而连接成线,从而在半导体阵列内部构建金属的三维网络结构[244]。相比于单金属颗粒,三维网络结构有利于实现载流子密度不同的半导体纳米棒之间的光生电子共享,半导体纳米棒由于自身尺寸及所处环境的不同,其内部光生载流子密度也不尽相同,光生电子-空穴对在高载流子密度的纳米棒中由于不能及时导走而更易于复合,而三维网络结构可以有效缓解这一问题,使载流子密度高的纳米棒中的光生电子可以通过相邻载流子密度低的纳米棒快速导走,从而提升电子与空穴的分离效率,提高半导体阵列的光催化性能。

Jiao 等人首先利用气相沉积法将金属 Bi 沉积于 TiO_2 纳米管上,金属 Bi 表层可以被空气中的 O_2 氧化形成 Bi_2O_3 保护层,在原位 X 射线光电子能谱(XPS)的测试过程中,发现 Bi 层表面的 Bi_2O_3 会被光生电子逐渐还原为金属 Bi,但无法推断光生电子从何而来,又如何转移到 Bi_2O_3 层上而将其还原。为了对其光生电子的具体转移路径进行研究,对 Bi、Bi_2O_3、Bi/TiO_2 和 Bi_2O_3/TiO_2 四种不同的样品分别在紫外光和可见光照射下的同步光照-XPS 进行了测试(图 1-65),结果发现 Bi_2O_3 无论在紫外光还是可见光下都不会发生还原反应,从而得知 Bi_2O_3 在光照条件下是稳定的,还原电子并非来自于 Bi_2O_3 本身。而 Bi_2O_3/TiO_2 在可见光下不发生还原反应,但在紫外光照射时却可以发生,由此可知来自于 TiO_2 的电子

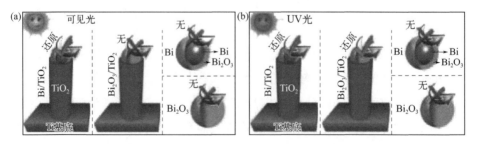

图 1-65 同步光照-XPS 结果推导可见光下光生电荷具体转移路径图

可以将 Bi_2O_3 还原。另外，表面有 Bi_2O_3 层的金属铋在单独照射时也不发生还原反应，但将其负载到 TiO_2 上之后（图 1-66），无论是可见光还是紫外光照射都可以发生还原反应，而在可见光照射下 TiO_2 的价电子是不能被激发的，由此可以推断在可见光照射下发生还原反应的光生电子来源于金属 Bi 的表面等离子体共振效应，但它却不能直接从 Bi 转移到 Bi_2O_3 上（因为带有氧化层的金属铋不发生还原反应），而是先迁移到 TiO_2 的导带然后再转移到 Bi_2O_3 上

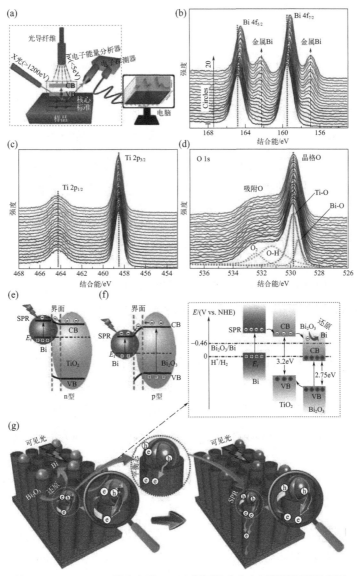

图 1-66　Bi/TiO_2 同步光照-XPS 谱图及光生电子转移机理示意图

将其还原，这一发现首次从微观上揭示了 Bi/TiO_2 光生电荷具体的转移路径，弄清了金属铋/半导体复合结构在光催化过程中的具体机理，同时也从实验方面进一步证实了金属铋在可见光区具有显著的表面等离子体共振效应，从而证明基于金属铋的表面等离子体共振效应构建高效的金属铋/半导体复合材料是切实可行的，采用价格低廉的金属铋作为贵金属金和银的替代者开展广泛的科学研究具有重要学术价值和广阔应用前景[245]，此外，金属铋表面等离子体共振效应产生的电子会优先转移到 TiO_2 的导带而不是直接转移到 Bi_2O_3 表面的原因也进行了理论分析，认为这是由于 TiO_2 和 Bi_2O_3 属于不同类型的半导体，金属铋分别与它们接触，表面能带结构会向不同方向发生弯曲导致的。

1.4.2.2 铋系半导体/氧化物半导体异质结光催化剂

大部分铋系半导体吸收可见光，铋系半导体与其他氧化物半导体复合能拓宽光响应区域，也促使光生电子和空穴发生快速迁移，降低复合效率，提高光催化材料的催化活性。根据铋系半导体/氧化物半导体异质结中的光生电子和空穴传输路径，异质结分为：Ⅱ型异质结和 Z 型异质结。两个能带交错排列的半导体组成Ⅱ型异质结。光照下光生电子从导带位较高的半导体迁移到较低的半导体，光生空穴从价带位较低的半导体迁移到较高的半导体，光生载流子在界面处快速分离和迁移[246-250]。如图 1-67（a）所示，Li 等研究者制备了 BiOCl/TiO_2 复合光催化剂，选择罗丹明 B 的降解来检测其催化性能[251]。在光激发下，TiO_2 中的光生电子将转移到 BiOCl 的导带中与 O_2 生成·O_2^-，与罗丹明 B 发生反应；BiOCl 中的光生空穴将转移到 TiO_2 的价带中，与罗丹明 B 发生氧化反应，异质结有效地提升了光生载流子的分离效率。实验结果表明：BiOCl/TiO_2 复合光催化材料的活性比 BiOCl 和 TiO_2 都高。如图 1-67(b) 所示，Wang 等研究者制备了 BiOBr/TiO_2 纳米纤维光催化剂，在光激发下，BiOBr 的光生电子转移到 TiO_2 导带中与吸附的 O_2 反应生成·O_2^-；TiO_2 的空穴转移到 BiOBr 的价带氧化染料分子。BiOBr/TiO_2 纳米纤维光催化剂相比于 TiO_2 纳米纤维具有更强的光催化活性[252]。Jiao 等人制备了 Mo 掺杂 $BiVO_4$、Bi_2MoO_6/ZnO 等多种半导体纳米材料，采用同步光照-XPS 技术发现其在开光/闭光条件下具有不同的电子转移方向，从而推断出了光生电子在不同元素之间的转移机制，为探索其光催化反应机理提供重要依据（图 1-68）。

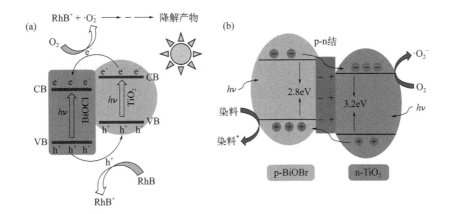

图1-67　BiOCl/TiO$_2$异质结降解罗丹明B的机理图（a）和BiOBr/TiO$_2$异质结纳米纤维降解有机物污染物的机理图（b）

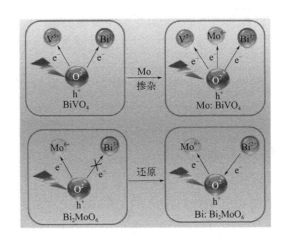

图1-68　Bi$_2$MoO$_6$/ZnO同步光照-XPS谱图及光生电子转移机理示意图

科学家根据绿色植物中电子的迁移路径，制备了Z型异质结光催化材料[253]。Z型异质结分为两类：有导电介质的Z型异质结和无导电介质的Z型异质结。有导电介质的Z型异质结：较低导带位中的光生电子沿着媒介迁移到较高价带位中与空穴复合，从而留下有还原能力的电子和氧化能力的空穴[254-256]。如图1-69（a）所示，Wan等人制备了Co$_3$O$_4$/Ag/Bi$_2$WO$_6$ Z型异质结的光催化复合材料，并研究了其去除重金属离子Cr（Ⅵ）和盐酸四环素（TCH）的光催化活性。研究结果表明：复合光催化剂具有光谱响应宽、氧化还原能力强等优点，其光催化活性明显优于单组分的催化剂[257]。但是这种Z

型异质结中的导电介质贵金属（Au、Pt等）增加了材料的合成难度和成本。无导电介质的Z型异质结：较低导带位中的光生电子与较高价带位中的空穴发生复合，不需要导电介质的传输[258-260]。Chen等人制备了WO₃/Bi₂WO₆ Z型异质结，研究了其光催化降解水杨酸的活性［图1-69（b）］。研究结果表明：Z型异质结的光电流强度明显高于单组分的，这说明构建的Z型异质结能有效分离光生载流子，提高光催化活性[261]。

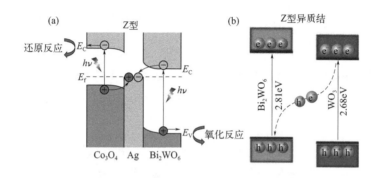

图1-69　Co₃O₄/Ag/Bi₂WO₆ Z型异质结的光催化机理图（a）和WO₃/Bi₂WO₆ Z型异质结光生载流子分离转移示意图（b）

1.4.3　铋系光催化材料制备方法

铋系光催化材料的制备方法根据制备状态的不同可分为：固相法、气相法和液相法[262-264]。固相法和气相法工艺复杂、反应时间长、对设备要求高与成本高等劣势不利于实现工业化生产；液相法反应设备简单，反应易于控制，反应物纯度高，是目前广泛采用的合成方法。所以，本文主要简单介绍铋系光催化材料的三种主要的液相法：水热或溶剂热法、溶胶-凝胶法、静电纺丝技术。

1.4.3.1　水热或溶剂热法

水热或溶剂热法是指在密闭的高温、高压的空间内，以水或有机物为溶剂，使前驱体反应物发生化学反应的方法（图1-70）。通过改变反应体系的温度、时间、升温速率和前驱体配比等参数，可以获得不同形貌和结构的产物[265-267]。水热或溶剂热法具有合成步骤简单、易于大规模化而且产物种类多、晶型好、纯度高、污染小等优点，是科研工作者首选的制备方法，也极具工业应用前景。如图 1-71 所示,Liu 等人采用水热法制备复合光催化剂 Bi_2O_4/Bi_2WO_6,研究结果表明:复合光催化材料 Bi_2O_4/Bi_2WO_6 光催化活性相比单组分 Bi_2O_4 和 Bi_2WO_6 有较大的提高[268]。

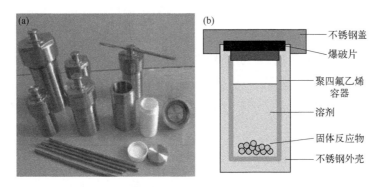

图 1-70　高压反应釜的实物图(a)及示意图(b)

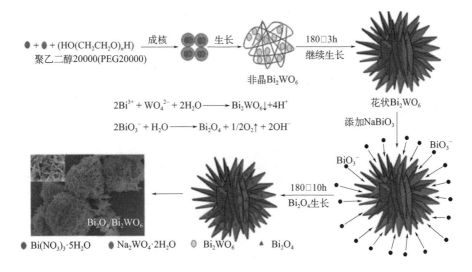

图 1-71　Bi_2O_4/Bi_2WO_6 复合光催化剂的生长机理示意图

1.4.3.2　溶胶-凝胶法

溶胶-凝胶法:反应物的醇盐或有机络合物等水解生成相应的氢氧化物或含水氧化物等形式的溶胶,然后经过一步缩合聚合而形成凝胶固化,最后分离、干燥。其优点是纯度高、颗粒分布均匀且尺度小、反应温度低、过程容易控制等。但原材料成本较高,步骤较多,不是特别适用于大规模工业生产。

1.4.3.3 静电纺丝技术

通过调节表面活性剂、结构导向剂等实验条件，采用溶胶-凝胶等方法可制备出一维结构的纳米纤维。但是这些方法操作复杂，产物形态不稳定。静电纺丝法是制备超长一维纳米纤维材料的高效、简单的方法。如图 1-72 所示，先在装置容器中装入适量的纺丝溶液，再在喷丝头端施加高压电，纺丝液滴在电场力的作用下喷出细流，形成泰勒锥。高分子发生固化，电纺纤维被接收板收集，形成交错的网毡结构，经过热处理后形成不同化学组成的无机纳米纤维材料[269-271]。纤维具有较高的比表面积，在复合纳米材料中作基底材料同时为其二次生长过程提供反应活性位点，有利于次级结构的均匀生长；另外纤维的宏观网毡结构有利于催化剂的分离与回收[272]。

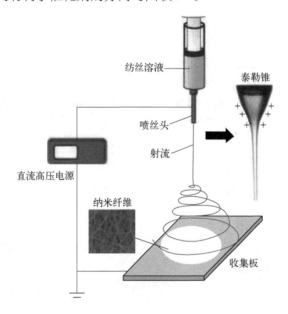

图 1-72　静电纺丝装置示意图

1.5　主要研究内容

铋系半导体材料因其在可见光下具有良好的催化作用而成为新型光催化材料的研究热点之一。铋系光催化剂以其独特的电子结构、优良的可见光吸收能力和较高的有机物降解能力，在光电化学（PEC）分解水、降解有机物以及光还原 CO_2 等方面都有广泛的应用。MOFs 作为一种对光敏感的半导体材料，其中的有机配体部分起到捕光天线分子的作用，然后将吸收的光子高效地传输给 MOFs 中的金属团簇，同时 MOFs 的比表面积大、孔径大小可调，构成 MOFs 的金属离子（或团簇）和有机多齿配体丰富多样且具有可剪裁性，是性能优良的载体，MOFs 材料的这些优点使其在选择性吸附分离、催化等领域具有非常广阔的应用前景。因此，实现两种材料在特点及优点上的结合，充分发挥两者之间的电子传输性能与催化协同作用，有望获得较好的光激发电子和空穴分离效果，以获得更高的光利用效率、光电转换效率和光催化活性。具体研究如下。

1.5.1　Bi_2WO_6/MIL-100（Fe）复合光催化材料

采用水热法制备了 MIL-100（Fe），然后采用原位复合法制备了 Bi_2WO_6/MIL-100（Fe）复合光催化材料。并对复合光催化纳米材料降解有机污染物（如水杨酸）进行了详细的研究，实验结果发现 Bi_2WO_6/MIL-100（Fe）具有比纳米 Bi_2WO_6 颗粒和 MIL-100（Fe）更高的光催化性能。n 型 Bi_2WO_6 与 p 型 MIL-100（Fe）形成 p-n 结可以有效地提高电子 - 空穴的分离效率，并有利于电子的传导，从而可以显著地增强半导体纳米材料的光催化及光电转换性能。

1.5.2　Fe/W 共掺杂 $BiVO_4$/MIL-100（Fe）复合光催化材料

金属有机框架结构（MOFs）在光催化中的应用成为近几年来的研究热点，然而关于 MOFs 作为助催化剂在光电转换中的应用却鲜见报道，首次将 MOFs 结构应用于 $BiVO_4$ 光电极的光催化反应，发现 MOFs 具有极好的助催化性能，可以大幅度提高 $BiVO_4$ 的光电转换效率。此外，尽管 $BiVO_4$ 的化学性质稳定，但是其作为光电极在光催化中光电流会随着时间延长而减小，我们通过少量 Fe 离子掺杂的方法有效地解决了 $BiVO_4$ 的光电转换稳定性差的问题，分析认为采用少量 Fe 离子替代部分 Bi 离子可以有效地减少 $BiVO_4$ 形成过程中的表面缺陷，使其在持续光照条件下具有极高的光化学稳定性，并且 Fe 离子掺杂可以与其他提高光催化性能的方法形成协同效应，例如在利用 Fe 取代 Bi 的同时采用少量高价态的 W 替代低价态的 V，这种更高价态元素的掺杂是一种典型的 n 型掺杂，可以有效地增加半导体内部光生载流子的密度，从而在保持 $BiVO_4$ 光电转换稳定性的同时，进一步提高其光电转换效率，因此本工作不但从实验角度解决了 $BiVO_4$ 光电性能低、稳定性差的问题，更为其将来的实际工业化应用奠定了物质基础。

1.5.3 UiO-66/BiOI 复合光催化材料

采用水热法制备了 UiO-66，然后采用原位复合法制备了 UiO-66/BiOI 复合光催化材料。并对复合光催化纳米材料降解有机污染物（如对硝基苯酚）进行了详细的研究，实验结果发现 UiO-66/BiOI 具有比纳米 BiOI 颗粒和 UiO-66 更高的光催化性能。n 型 UiO-66 与 p 型 BiOI 形成 p-n 结可以有效地提高电子 - 空穴的分离效率，并有利于电子的传导，从而可以显著地增强半导体纳米材料的光催化及光电转换性能。

1.5.4 磁性树脂/BiOI 复合光催化材料

聚丙烯酸阴离子树脂是一种新型离子交换树脂，有良好的大孔网状结构和较大的比表面积，将 BiOI 负载在磁性聚丙烯酸阴离子树脂表面，制备磁性树脂/BiOI 复合光催化材料降解 H 酸。实验结果表明磁性树脂/BiOI 具有比纳米 BiOI 颗粒更好的光催化性能。BiOI 与树脂中的 Fe_3O_4 形成异质结可以有效地提高电子 - 空穴的分离效率，从而可以显著地增强半导体纳米材料的光催化及光电转换性能。

1.5.5 $FeVO_4/C_3N_4$ 复合光催化材料

制备 $FeVO_4$ 和 C_3N_4 异质结，研究发现，超薄的 $g-C_3N_4$ 片层能够有效地捕捉空穴，有利于异质结的光生载流子的分离，提高了光电流密度，从而提高了 $FeVO_4$ 的光电化学性能。

1.5.6 Fe 掺杂 $BiVO_4$ 与 MIL-53(Fe) 复合光催化材料

将 MIL-53(Fe) 与 Fe 离子掺杂后的 $BiVO_4$ 复合，提高了 $BiVO_4$ 的光电稳定性及光电催化反应活性。

参考文献

[1] Baur E, Perret A. Uber die einwirkung von licht auf geloste silbersalze in gegenwart von

zinkoxyd[J]. Helvetica Chimica Acta, 1924, 7(1): 910-915.

[2] Fujishima A, Honda K. Electrochemical photolysis of water at a semiconductor electrode[J]. Nature, 1972, 238: 37-38.

[3] Singh P, Borthakur A. A review on biodegradation and photocatalytic degradation of organic pollutants: A bibliometric and comparative analysis[J]. Journal of Cleaner Production, 2018, 196: 1669-1680.

[4] Zhang G, Chen D, Li N, et al. Fabrication of Bi_2MoO_6/ZnO hierarchical heterostructures with enhanced visible-light photocatalytic activity[J]. Appl Catal B Environ, 2019, 250: 313-324.

[5] Wang J, Tang L, Zeng G, et al. 0D/2D interface engineering of carbon quantum dots modified Bi_2WO_6 ultrathin nanosheets with enhanced photoactivity for full spectrum light utilization and mechanism insight[J]. Appl Catal B Environ, 2018, 222: 115-123.

[6] Zhang J, Niu C, Ke J, et al. Ag/AgCl/Bi_2MoO_6 composite nanosheets: A plasmonic Z-scheme visible light photocatalyst[J]. Catal Commun, 2015, 59: 30-34.

[7] Di J, Chen C, Zhu C, et al. Bismuth vacancy mediated single unit cell Bi_2WO_6 nanosheets for boosting photocatalytic oxygen evolution[J]. Appl Catal B Environ, 2018, 238: 119-125.

[8] Li C, Chen G, Sun J, et al. A novel mesoporous single-crystal-like Bi_2WO_6 with enhanced photocatalytic activity for pollutants degradation and oxygen production[J]. ACS Appl Mater Inter, 2015, 7(46): 25716-25724.

[9] Zhu Z, Li Y, Wang C, et al. Facile synthesis and characterization of Bi_2MoO_6/Ag_3PO_4/RGO composites with enhanced visible-light-driven photocatalytic activity[J]. Mater Lett, 2018, 227: 296-300.

[10] Liu Y, Jia J, Li Y, et al. Novel ZnO/NiO janus-like nanofibers for effective photocatalytic degradation[J]. Nanotechnology, 2018, 29: 435704.

[11] Pardeep S, Anwesha B. A review on biodegradation and photocatalytic degradation of organic pollutants: A bibliometric and comparative analysis[J]. Journal of Cleaner Production, 2018, 196: 1669-1680.

[12] Kumaravel V, Mathew S, Bartlett J, et al. Photocatalytic hydrogen production using metal doped TiO_2: A review of recent advances[J]. Applied Catalysis B: Environmental, 2019, 244: 1021-1064.

[13] Liu Z, Sun D, Guo P, et al. An efficient bicomponent TiO_2/SnO_2 nanofiber photocatalyst fabricated by electrospinning with a side-by-side dual spinneret method[J]. Nano Lett, 2007, 7: 1081-1085.

[14] Liu H, Han C, Shao C, et al. ZnO/$ZnFe_2O_4$ janus hollow nanofibers with magnetic separability for photocatalytic degradation of water-soluble organic dyes[J]. ACS Appl Nano Mater, 2019, 2(8): 4879-4890.

[15] Wang W, Li N, Chi Y, et al. Electrospinning of magnetical bismuth ferrite nanofibers with photocatalytic activity[J]. Ceram Int, 2013, 39(4): 3511-3518.

[16] 田月峰. 钒酸铋光催化复合材料的制备及对大肠杆菌灭活性能研究[D]. 呼和浩特: 内蒙古大学, 2019.

[17] Nakata K, Fujishima A. TiO_2 photocatalysis: Design and applications[J]. Journal of Photochemistry and Photobiology C: Photochemistry Reviews, 2012, 13(3): 169-189.

[18] Zhu Z, Iyemperumal S, Kushnir K, et al. Enhancing the solar energy conversion efficiency of solution-deposited Bi_2S_3 thin films by annealing in sulfur vapor at elevated temperature[J]. Sustainable Energy & Fuels, 2017, 1: 2134-2144.

[19] Du H, Liu Y, Shen C, et al. Nanoheterostructured photocatalysts for improving photocatalytic hydrogen production[J]. Chinese Journal of Catalysis, 2017, 38(8): 1295-1306.

[20] Tokunaga S, Kato H, Kudo A. Selective preparation of monoclinic and tetragonal $BiVO_4$ with scheelite structure and their photocatalytic properties[J]. Chemistry of Materials, 2001, 13(12): 4624-4628.

[21] Zhao Y, Li R, Mu L, et al. The significance of crystal morphology controlling in semiconductor-based photocatalysis: A case study on $BiVO_4$ photocatalyst[J]. Crystal Growth & Design, 2017, 17(6): 2923-2928.

[22] Xu Y, He X, Zhong H, et al. Solid salt confinement effect: An effective strategy to fabricate high crystalline polymer carbon nitride for enhanced photocatalytic hydrogen evolution[J]. Applied Catalysis B: Environmental, 2019, 246: 349-355.

[23] Wang W, Li G, Xia D, et al. Photocatalytic nanomaterials for solar-driven bacterial inactivation: Recent progress and challenges[J]. Environmental Science: Nano, 2017, 4(4): 782-799.

[24] Zhang X, Zhao X, Wu D, et al. $MnPSe_3$ monolayer: A promising 2D visible-light photohydrolytic catalyst with high carrier mobility[J]. Advanced Science, 2016, 3(10): 1600062.

[25] Li X, Dai Y, Li M, et al. Stable Si-based pentagonal monolayes: High carrier mobilities and applications in photocatalytic water splitting[J]. Journal of Materials Chemistry A, 2015, 3: 24055-24063.

[26] Li X, Xiong J, Gao X, et al. Recent advances in 3D g-C_3N_4 composite photocatalysts for photocatalytic water splitting, degradation of pollutants and CO_2 reduction[J]. Journal of Alloys and Compounds, 2019, 802: 196-209.

[27] Du J, Ma S, Yan Y, et al. Corn-silk-templated synthesis of TiO_2 nanotube arrays with Ag_3PO_4 nanoparticles for efficient oxidation of organic pollutants and pathogenic bacteria under solar light[J]. Colloids and Surfaces A: Physicochemical and Engineering Aspects, 2019, 572: 237-249.

[28] Midya L, Patra A, Banerjee C, et al. Novel nanocomposite derived from ZnO/CdS QDs embedded crosslinked chitosan: An efficient photocatalyst and effective antibacterial agent[J]. Journal of Hazardous Materials, 2019, 369: 398-407.

[29] Yu Y, Ding Y, Zuo S, et al. Photocatalytic activity of nanosized cadmium sulfides synthesized by complex compound thermolysis[J]. International Journal of Photoenergy, 2011, 2011: 1-5.

[30] Yang X, Qin J, Jiang Y, et al. Fabrication of P25/Ag$_3$PO$_4$/graphene oxide heterostructures for enhanced solar photocatalytic degradation of organic pollutants and bacteria[J]. Applied Catalysis B: Environmental, 2015, 166-167: 231-240.

[31] Liu G, Yu J, Lu G, et al. Crystal facet engineering of semiconductor photocatalysts: Motivations, advances and unique properties[J]. Chemical Communications, 2011, 47(24): 6763-6783.

[32] Martin D, Umezawa N, Chen X, et al. Facet engineered Ag$_3$PO$_4$ for efficient water photooxidation[J]. Energy & Environmental Science, 2013, 6(11): 3380-3386.

[33] Pan J, Liu G, Lu G, et al. On the true photoreactivity order of {001}, {010}, and {101} facets of anatase TiO$_2$ crystals[J]. Angewandte Chemie International Edition, 2011, 50(9): 2133-2137.

[34] Xie Y, Liu G, Yin L, et al. Crystal facet-dependent photocatalytic oxidation and reduction reactivity of monoclinic WO$_3$ for solar energy conversion[J]. Journal of Materials Chemistry, 2012, 22(14): 6746-6751.

[35] Jiao Z, Zhang Y, Yu H, et al. Concave trisoctahedral Ag$_3$PO$_4$ microcrystals with high-index facets and enhanced photocatalytic properties[J].Chem Commum, 2013, 49: 636-638.

[36] Wang D, Jiang H, Zong X, et al. Crystal facet dependence of water oxidation on BiVO$_4$ sheets under visible light irradiation[J]. Chemistry - A European Journal, 2011, 17(4): 1275-1282.

[37] Liu G, Wang L, Yang H, et al. Titania-based photocatalysts-crystal growth, doping and heterostructuring[J]. Journal of Materials Chemistry, 2010, 20(5): 831-843.

[38] Jiao Z, Chen T, Xiong J, et al.Visible-light-driven photoelectrochemical and photocatalytic performances of Cr-doped SrTiO$_3$/TiO$_2$ heterostructured nanotube arrays[J].Sci Rep, 2013, 3: 2720-2728.

[39] Jiao Z, Zhang Y, Chen T, et al. TiO$_2$ nanotube arrays modified with Cr-doped SrTiO$_3$ nanocubes for highly efficient hydrogen evolution under visible light[J]. Chem Eur J, 2014, 20: 2654.

[40] Jiao Z, Chen T, Yu H, et al. Morphology modulation of SrTiO$_3$/TiO$_2$ heterostructures for enhanced photoelectrochemical performance[J]. J Colloid Inter Sci, 2014, 419: 95.

[41] Jiao Z, Zhang Y, Ouyang S, et al. BiAg alloy nanospheres: A new photocatalyst for H$_2$evolution from water splitting[J]. ACS Appl MaterInterfaces, 2014, 6: 19488.

[42] Rao H, Lim C, Bonin J, et al. Visible-light-driven conversion of CO$_2$ to CH$_4$ with an organic sensitizer and an iron porphyrin catalyst[J]. Journal of the American Chemical Society, 2018, 140(51): 17830-17834.

[43] Schwartzberg A, Zhang J. Novel optical properties and emerging applications of metal nanostructures[J].J Phys Chem C, 2008, 112(28): 10323-10337.

[44] Awazu K, Fujimaki M, Rockstuhl C, et al. A plasmonic photocatalyst consisting of silver nanoparticles embedded in titanium dioxide[J]. J Am Chem Soc, 2008, 130(5): 1676-1680.

[45] Wang P, Huang B, Zhang Q, et al. Highly efficient visible light plasmonic photocatalyst Ag@Ag(Br, I)[J]. Chem Eur J, 2012, 16(33): 10042-10047.

[46] Wang P, Huang B, Zhang X, et al.Highly efficient visible-light plasmonic photocatalyst Ag@AgBr[J].Chem Eur J, 2009, 15(8): 1821-1824.

[47] Wang P, Huang B, Qin X, et al. Ag@AgCl: A highly efficient and stable photocatalyst active under visible light[J].Angew Chem Int Ed, 2008, 47(41): 7931-7933.

[48] Gomes Silva C, Juárez R, MarinoT, et al. Influence of excitation wavelength (UV or visible light) on the photocatalytic activity of titania containing gold nanoparticles for the generation of hydrogen or oxygen from water[J].J Am Chem Soc, 2010, 133(3): 595-602.

[49] Chen J, Wu J, Wu P, et al. Plasmonic photocatalyst for H_2 evolution in photocatalytic water splitting[J]. J Phys Chem C, 2010, 115(1): 210-216.

[50] Furube A, Du L, Hara K, et al. Ultrafast plasmon-induced electron transfer from gold nanodots into TiO_2 nanoparticles[J]. J Am Chem Soc, 2007, 129(48): 14852-14853.

[51] Mubeen S, Hernandez-Sosa G, Moses D, et al. Plasmonic photosensitization of a wide band gap semiconductor: Converting plasmons to charge carriers[J]. Nano letters, 2011, 11(12): 5548-5552.

[52] Rhodes C, Franzen S, Maria J, et al. Surface plasmon resonance in conducting metal oxides[J].J Appl Phys, 2006, 100(5): 054905.

[53] Buonsanti R, Llordes A, Aloni S, et al. Tunable infrared absorption and visible transparency of colloidal aluminum-doped zinc oxide nanocrystals[J]. Nano letters, 2011, 11(11): 4706-4710.

[54] Shahzad M, Medhi G, Peale R, et al. Infrared surface plasmons on heavily doped silicon[J]. J Appl Phys, 2011, 110(12), 123105.

[55] Xing M, Zhang J, Chen F, et al. An economic method to prepare vacuum activated photocatalysts with high photo-activities and photosensitivities[J]. Chem Comm, 2011, 47(17): 4947-4949.

[56] Xing M, Fang W, Nasir M, et al. Self-doped Ti^{3+} enhanced TiO_2 nanoparticles with a high-performance photocatalysis[J]. Journal of Catalysis, 2013, 297: 236-243.

[57] Scotognella F, Della Valle G, Srimath Kandada A, et al.Plasmon dynamics in colloidal $Cu_{2-x}Se$ nanocrystals[J]. Nano letters, 2011, 11(11): 4711-4717.

[58] Zhao Y, Pan H, Lou Y, et al. Plasmonic $Cu_{2-x}S$ nanocrystals: Optical and structural properties of copper-deficient copper (I) sulfides[J]. J Am Chem Soc, 2009, 131(12): 4253-4261.

[59] Gordon T, Cargnello M, Paik T, et al. Nonaqueous synthesis of TiO_2 nanocrystals using TiF_4 to engineer morphology, oxygen vacancy concentration, and photocatalytic activity[J]. J Am Chem Soc, 2012, 134(15): 6751-6761.

[60] Manthiram K, Alivisatos A. Tunable localized surface plasmon resonances in tungsten oxide nanocrystals[J].J Am Chem Soc, 2012, 134(9): 3995-3998.

[61] Cheng H, Kamegawa T, Mori K, et al. Surfactant-free nonaqueous synthesis of plasmonic molybdenum oxide nanosheets with enhanced catalytic activity for hydrogen generation from ammonia borane under visible light[J].Angew Chem Int Ed, 2014, 126(11): 2954-2958.

[62] Justicia I, Ordejón P, Canto G, et al. Designed self-doped titanium oxide thin films for efficient visible-light photocatalysis[J].Adv Mater, 2002, 14(19), 1399-1402.

[63] Wang G, Ling Y, Wang H, et al. Hydrogen-treated WO_3 nanoflakes show enhanced photostability[J]. Energy & Environmental Science, 2012, 5(3): 6180-6187.

[64] Gan J, Lu X, Wu J, et al. Oxygen vacancies promoting photoelectrochemical performance of In_2O_3 nanocubes[J]. Scientific reports, 2013, 3: 1021.

[65] Wang J, Wang Z, Huang B, et al.Oxygen vacancy induced band-gap narrowing and enhanced visible light photocatalytic activity of ZnO[J]. ACS Appl Mater Interfaces, 2012, 4(8): 4024-4030.

[66] Zou X, Wang Y, et al. Direct conversion of urea into graphitic carbon nitride over mesoporous TiO_2 spheres under mild condition[J]. Chem Commun, 2011, 47(3): 1066-1068.

[67] Cao F, Wang J, Li S, et al. Rapid room-temperature synthesis and visible-light photocatalytic properties of BiOI nanoflowers[J]. J Alloy Compoud, 2015, 639: 445-451.

[68] Kong J, Rui Z, Wang X, et al. Visible-light decomposition of gaseous toluene over $BiFeO_3$-$(Bi/Fe)_2O_3$ heterojunctions with enhanced performance[J]. Chem Eng J, 2016, 302: 552-559.

[69] Xu Q, Cheng S, Hao X, et al. Effect of Ag doping on the formation and properties of percolative $Ag/BiFeO_3$ composite thin film by sol-gel method[J]. Appl Phys A Mater, 2017, 123(4): 1-12.

[70] Chaiwichian S, Wetchakun K, Kangwansupamonkon W, et al. Novel visible-light-driven $BiFeO_3$-Bi_2WO_6 nanocomposites toward degradation of dyes[J]. J Photoch Photobio A, 2017, 349: 183-192.

[71] Wang K, Shao C, Li X, et al. Hierarchical heterostructures of p-type BiOCl nanosheets on electrospun n-type TiO_2 nanofibers with enhanced photocatalytic activity[J]. Catal Commun, 2015, 67: 6-10.

[72] Zhang M, Shao C, Mu J, et al. One-dimensional Bi_2MoO_6/TiO_2 hierarchical heterostructures with enhanced photocatalytic activity[J]. Cryst Eng Comm, 2012, 14: 605-612.

[73] Jia X, Cao J, Lin H, et al. Transforming type-I to type-II heterostructure photocatalyst via energy band engineering: A case study of I-BiOCl/I-BiOBr[J]. Appl Catal B Environ, 2017, 204: 505-514.

[74] Weng S, Chen B, Xie L, et al. Facile in situ synthesis of a Bi/BiOCl nanocomposite with high photocatalytic activity[J]. J Mater Chem A, 2013, 1(9): 3068-3075.

[75] Wang L, Niu C, Wang Y, et al. The synthesis of $Ag/AgCl/BiFeO_3$ photocatalyst with enhanced visible photocatalytic activity[J]. Ceram Int, 2016, 42(16): 18605-18611.

[76] Liu W, Qiao L, Zhu A, et al. Constructing 2D BiOCl/C$_3$N$_4$ layered composite with large contact surface for visible-light-driven photocatalytic degradation[J]. Appl Surf Sci, 2017, 426: 897-905.

[77] Diesen V, Jonsson M. Formation of H$_2$O$_2$ in TiO$_2$ photocatalysis of oxygenated and deoxygenated aqueous systems: A probe for photocatalytically produced hydroxyl radicals[J]. J Phys Chem C, 2014, 118(19): 10083-10087.

[78] Xiao X, Xing C, He G, et al. Solvothermal synthesis of novel hierarchical Bi$_4$O$_5$I$_2$ nanoflakes with highly visible light photocatalytic performance for the degradation of 4-tert-butylphenol[J]. Appl Catal B Environ, 2014, 148-149: 154-163.

[79] Köferstein R, Buttlar T, Ebbinghaus S. Investigations on Bi$_{25}$FeO$_{40}$ powders synthesized by hydrothermal and combustion-like processes[J]. J Solid State Chem, 2014, 217: 50-56.

[80] Xiong Z, Cao L. Tailoring morphology, enhancing magnetization and photocatalytic activity via Cr doping in Bi$_{25}$FeO$_{40}$[J]. J Alloy Compd, 2019, 773: 828-837.

[81] He Y, Zhang L, Wang X, et al. Enhanced photodegradation activity of methyl orange over Z-scheme type MoO$_3$-g-C$_3$N$_4$ composite under visible light irradiation[J]. RSC Adv, 2014, 4(26): 13610-13619.

[82] Huang S, Wang H, Zhu N, et al. Metal recovery based magnetite near-infrared photocatalyst with broadband spectrum utilization property[J]. Appl Catal B Environ, 2016, 181: 456-464.

[83] Zhang J, Niu C, Ke J, et al. Ag/AgCl/Bi$_2$MoO$_6$ composite nanosheets: A plasmonic Z-scheme visible light photocatalyst[J]. Catal Commun, 2015, 59: 30-34.

[84] Yang X, Tian L, Zhao X, et al. Interfacial optimization of g-C$_3$N$_4$-based Z-scheme heterojunction toward synergistic enhancement of solar-driven photocatalytic oxygen evolution[J]. Appl Catal B Environ, 2019, 244: 240-249.

[85] Miyoshi A, Vequizo J, Nishioka S, et al. Nitrogen/fluorine-Codoped rutile titania as a stable oxygen-evolution photocatalyst for solar-driven Z-scheme water splitting[J]. Sustain Energy Fuels, 2018, 2(9): 2025-2035.

[86] Li J, Wang J, Zhang G, et al. Enhanced molecular oxygen activation of Ni^{2+}-doped BiO$_{2-x}$ nanosheets under UV, visible and near-infrared irradiation: Mechanism and DFT study[J]. Appl Catal B Environ, 2018, 234: 167-177.

[87] Wang K, Zhang G, Li J, et al. 0D/2D Z-scheme heterojunctions of bismuth tantalate quantum dots/ultrathin g-C$_3$N$_4$ nanosheets for highly efficient visible light photocatalytic degradation of antibiotics[J]. ACS Appl Mater Inter, 2017, 9(50): 43704-43715.

[88] Zhang R, Han Q, Li Y, et al. Fabrication of a Ag$_3$PO$_4$/reduced graphene oxide/BiOBr ternary photocatalyst for enhanced visible-light photocatalytic activity and stability[J]. Journal of Alloys and Compounds, 2019, 810: 151868.

[89] Cheng H, Huang B, Qin X, et al. A controlled anion exchange strategy to synthesize Bi$_2$S$_3$ nanocrystals/BiOCl hybrid architectures with efficient visible light photoactivity[J]. Chemical Communications, 2012, 48 (1): 97-99.

[90] Kong L, Ji Y, Dang Z, et al. g-C_3N_4 loading black phosphorus quantum dot for efficient and stable photocatalytic H_2 generation under visible light[J]. Adv Funct Mater, 2018, 28(22): 1800668.

[91] Yu W, Chen J, Shang T, et al. Direct Z-scheme g-C_3N_4/WO_3 photocatalyst with atomically defined junction for H_2 production[J]. Appl Catal B Environ, 2017, 219: 693-704.

[92] Lian Z, Xu P, Wang W, et al. C60-decorated CdS/TiO_2 mesoporous architectures with enhanced photostability and photocatalytic activity for H_2 evolution[J]. ACS Appl Mater Inter, 2015, 7(8): 4533-4540.

[93] Jin Z, Zhang Q, Chen J, et al. Hydrogen bonds in heterojunction photocatalysts for efficient charge transfer[J]. Appl Catal B Environ, 2018, 234: 198-205.

[94] Mao L, Cai X, Yang S, et al. Black phosphorus-CdS-$La_2Ti_2O_7$ ternary composite: Effective noble metal-free photocatalyst for full solar spectrum activated H_2 production[J]. Appl Catal B Environ, 2019, 242: 441-448.

[95] Zhao J, Li Y, Liu P, et al. Local coulomb attraction for enhanced H_2 evolution stability of metal sulfide photocatalysts[J]. Appl Catal B Environ, 2018, 221: 152-157.

[96] Liu J, Xiong C, Jiang S, et al. Efficient evolution of reactive oxygen species over the coordinated π-delocalization g-C_3N_4 with favorable charge transfer for sustainable pollutant elimination[J]. Appl Catal B Environ, 2019, 249: 282-291.

[97] Shi X, Fujitsuka M, Kim S, et al. Faster electron injection and more active sites for efficient photocatalytic H_2 evolution in g-C_3N_4/MoS_2 hybrid[J]. Small, 2018, 14(11): 1703277.

[98] Hao R, Wang G, Tang H, et al. Template-free preparation of macro/mesoporous g-C_3N_4/TiO_2 heterojunction photocatalysts with enhanced visible light photocatalytic activity[J]. Applied Catalysis B: Environmental, 2016, 187: 47-58.

[99] Wang W, Yu Y, An T, et al. visible-light-driven photocatalytic inactivation of E.coli K-12 by bismuth vanadate nanotubes: bactericidal performanceand mechanism[J]. Environmental Science & Technology, 2012, 46(8): 4599-4606.

[100] Ganguly P, Byrne C, Breen A, et al. Antimicrobial activity of photocatalysts: Fundamentals, mechanisms, kinetics and recent advances[J]. Applied Catalysis B: Environmental, 2018, 225: 51-75.

[101] Matsunaga T, Tomoda R, Nakajima T, et al. Continuous-sterilization system that uses photosemiconductor powders[J]. Appl Environ Microb, 1988, 54(6): 1330-1333.

[102] Zeng X, Lan S, Lo I. Rapid disinfection of E.coli by a ternary $BiVO_4$/Ag/g-C_3N_4 compo-site under visible light: Photocatalytic mechanism and performance investigation in authentic sewage[J]. Environmental Science: Nano, 2019, 6: 610-623.

[103] Qiu P, Xu C, Zhou N, et al. Metal-free black phosphorus nanosheets-decorated graphitic carbon nitride nanosheets with C-P bonds for excellent photocatalytic nitrogen fixation[J]. Appl Catal B Environ, 2018, 221: 27-35.

[104] Dong G, Ho W, Wang C. Selective photocatalytic N_2 fixation dependent on g-C_3N_4 induced by nitrogen vacancies[J]. J Mater Chem A, 2015, 3(46): 23435-23441.

[105] Bi Y, Wang Y, Dong X, et al. Efficient solar-driven conversion of nitrogen to ammonia in pure water via hydrogenated bismuth oxybromide[J]. RSC Adv, 2018, 8(39): 21871-21878.

[106] Gao X, Wen Y, Qu D, et al. Interference effect of alcohol on nessler's reagent in photocatalytic nitrogen fixation[J]. ACS Sustain Chem Eng, 2018, 6(4): 5342-5348.

[107] Liu J, Li Y, Liu H, et al. Photo-thermal synergistically catalytic conversion of glycerol and carbon dioxide to glycerol carbonate over Au/ZnWO$_4$-ZnO catalysts[J]. Appl Catal B Environ, 2019, 244: 836-843.

[108] Zhou B, Song J, Zhou H, et al. Light-driven integration of the reduction of nitrobenzene to aniline and the transformation of glycerol in to valuable chemicals in water[J]. RSC Advances, 2015, 5: 36347-36352.

[109] Zhai G, Liu Y, Lei L, et al. Light-promoter CO_2 conversation from epoxides to cyclic carbonates at ambient conditions over a Bi-based metal-organic framework[J]. ACS Catalysis, 2021, 11: 1988-1994.

[110] Shi Q, Li Z, Chen L, et al. Synthesis of SPR Au/BiVO$_4$ quantum dot/rutile-TiO$_2$ nanorod array composites as efficient visible-light photocatalysts to convert CO_2 and mechanism insight[J]. Appl Catal B Environ, 2019, 244: 641-649.

[111] Halmann M. Photoelectrochemical reduction of aqueous carbon dioxide on p-type gallium phosphide in liquid junction solar cells characterisation of organic acids trapped in coals[J]. Nature, 1978, 275(14): 115-116.

[112] Cui Y, Li B, He H, et al.Metal-organic frameworks as platforms for functional materials[J]. Accounts of Chemical Research, 2016, 49(3): 483-493.

[113] Banerjee R, Phan A, Wang B, et al. High-throughput synthesis of zeolitic imidazolate frameworks and application to CO_2 capture[J]. Science, 2008, 319(5865): 939-943.

[114] Stone F, Graham W. Inorganic polymer[M]. New York: Academic Press, 1962.

[115] Buser H, Schwarzenbach D, Petter W, et al. The crystal structure of Prussian Blue: $Fe_4[Fe(CN)_6]_3 \cdot xH_2O$[J]. Inorg Chem, 1977, 16: 2704-2710.

[116] Stock N, Biswas S. Synthesis of metal-organic frameworks (MOFs): Routes to various MOF topologies, morphologies, and composites[J]. Chem Rev, 2012, 112(2): 933-969.

[117] Dhainaut J, Bonneau M, Ueoka R, et al. Formulation of metal-organic framework inks for the 3D printing of robust microporous solids toward high-pressure gas storage and separation[J]. ACS Appl Mater Interfaces, 2020, 12(9): 10983-10992.

[118] Li C, Dong S, Tang R, et al. Heteroatomic interface engineering in MOF-derived carbon heterostructures with built-in electric-field effects for high performance Al-ion batteries[J]. Energy Environ Sci, 2018, 11 (11): 3201-3211.

[119] Morris P, McPherson I, Edwards M, et al. Electric field-controlled synthesis and characterisation of single metal-organic-framework (MOF)nanoparticles[J]. Angew Chem Int Ed, 2020, 59 (44): 19696-19701.

[120] Hoskins B, Robson R. Infinite polymeric frameworks consisting of three dimensionally linked rod-like segments[J]. Journal of the American Chemical Society, 1994, 116: 1151-1152.

[121] Yaghi O, Li G, Li H. Selective binding and removal of guests in a microporous metal-organic framework[J]. Nature, 1995, 378: 703-706

[122] Long J, Yaghi O. The pervasive chemistry of metal-organic frameworks[J]. Chemical Society Reviews, 2009, 38: 1213-1214.

[123] Koh K, Wong-Foy A, Matzger A. A porous coordination copolymer with over 5000m^2/g BET surface area[J]. Journal of the American Chemical Society, 2009, 131: 4184-4185.

[124] Hou C, Xu Q. Metal-organic frameworks for energy[J]. Advanced Energy Materials, 2018, 9(23): 1801307.

[125] Eddaoudi M, Kim J, Rosi N, et al. Systematic design of pore size and functionality in isoreticular MOFs and their application in methane storage[J]. Science, 2002, 295: 469-472.

[126] Furukawa H, Cordova K, O'Keeffe M, et al. The chemistry and applications of metal-organic frameworks[J]. Science, 2013, 341(6149): 1230444.

[127] Li B, Wen H, Cui Y, et al. Emerging multifunctional metal-organic framework materials[J]. Advanced Materials, 2016, 28(40): 8819-8860.

[128] Zhang T, Jin Y, Shi Y, et al. Modulating photoelectronic performance of metal-organic frameworks for premium photocatalysis[J]. Coord Chem Rev, 2019, 380: 201-229.

[129] Lee J, Kapustin E, Pei X, et al. Architectural stabilization of a gold (III catalyst in metal-organic frameworks[J]. Chem, 2020, 6(1): 142-152.

[130] Li R, Zhang W, Zhou K. Metal-organic-framework based catalysts for photoreduction of CO_2 [J]. Adv Mater, 2018, 30(35): 1705512.

[131] Cui Y, Li B, He H, et al. Metal-organic frameworks as platforms for functional materials[J]. Accounts of Chemical Research, 2016, 49(3): 483-493.

[132] Wu M, Yang Y. Metal-organic framework (MOF)-based drug/cargo delivery and cancer therapy[J]. Advanced Materials, 2017, 29(23): 1606134.

[133] Masoomi M, Morsali A, Dhakshinamoorthy A, et al. Mixed-metal MOFs: Unique opportunities in metal-organic framework (MOF) functionality anddesign[J]. Angewandte Chemie-International Edition, 2019, 58(43): 15188-15205.

[134] Wang S, McGuirk C, d'AquinoA, et al. Metal-organic framework nanoparticles[J]. Advanced Materials, 2018, 30(37): 1800202.

[135] Om Y, O'keeffe M, Nw O, et al. Reticular synthesis and the design of new materials[J]. Nature, 2003, 423(6941): 705-714.

[136] Farha O, Eryazici I, Jeong N, et al. Metal-organic framework materials with ultrahigh surface areas: Is the sky the limit[J]. Journal of the American Chemical Society, 2012,

134(36): 15016-15021.

[137] Lomachenko K. Metal-organic frameworks: Structure, properties, methods of synthesis and characterization[J]. Russian Chemical Reviews Reviews on Current Topics in Chemistry, 2016, 85(3): 280-307.

[138] Shuai Y, Liang F, Wang K, et al. Stable metal-organic frameworks: Design, synthesis, and applications[J]. Advanced Materials, 2018, 30(37): 1704303.

[139] Devic T, Serre C. High valence 3p and transition metal based MOFs[J]. Chemical Society Reviews, 2014, 43(16): 6097-6115.

[140] Lebedev O, Millange F, Serre C, et al. First direct imaging of giant pores of the metal-organic framework MIL-101[J]. Chemistry of Materials, 2005, 17: 6525-6527.

[141] Li R, Yuan Y, Qiu L, et al. A rational self-sacrificing template route to metal-organic framework nanotubes and reversible vapor-phase detection of nitroaromatic explosives[J]. Small, 2012, 8: 225-230.

[142] Pichon A, James S. An array-based study of reactivity under solvent-free mechanochemical conditions-insights and trends[J]. Crystengcomm, 2008, 10: 1839-1847.

[143] Pichon A, Lazuen-Garay A, James S. Solvent-free synthesis of a microporous metal-organic framework[J]. Crystengcomm, 2006, 8: 211-214.

[144] Yuan W, Frisic T, Apperley D, et al. High reactivity of metal-organic frameworks under grinding conditions: Parallels with organic molecular materials[J]. Angewandte Chemie-International Edition, 2010, 49: 3916-3919.

[145] Biemmi E, Christian S, Stock N, et al. High-throughput screening of synthesis paremeters in the formation of the metal-organic frameworks MOF-5 and HKUST-1[J]. Microporous and Mescroporous Materials, 2009, 117: 111-117.

[146] Chatel G, Colmenares J. Sonochemistry: From basic principles to innovative applications[J]. Topics in Current Chemistry, 2017, 375: 8.

[147] Zeng L, Guo X, He C, et al.Metal-organic frameworks: Versatile materials for heterogeneous photocatalysis[J]. ACS Catal, 2016, 6(11): 7935-7947.

[148] Hu Z, Deibert B, Li J. Luminescent metal-organic frameworks for chemical sensing and explosive detection[J]. Chem Soc Rev, 2014, 43(16): 5815-5840.

[149] Banerjee R, Phan A, Wang B, et al. High-throughput synthesis of zeolitic imidazolate frameworks and application to CO_2 capture[J]. Science, 2008, 319 (5865): 939-943.

[150] Yuan S, Feng L, Wang K, et al. Stable metal-organic frameworks: Design, synthesis, and applications[J]. Adv Mater, 2018, 30(37): e1704303.

[151] Lin R, Xiang S, Li B, et al. Our journey of developing multifunctional metal-organic frameworks[J]. Coord Chem Rev, 2019, 384: 21-36.

[152] Wu Y, Huang Z, Jiang H, et al. Facile synthesis of uniform metal carbide nanoparticles

[153] Zarekarizi F, Joharian M, Morsali A.Pillar-layered MOFs: Functionality, interpenetration, flexibility and applications[J]. J Mater Chem A, 2018, 6(40): 19288-19329.

[154] Xiao Y, Guo X, Yang N, et al. Heterostructured MOFs photocatalysts for water splitting to produce hydrogen[J]. Journal of Energy Chemistry, 2021, 58: 508-522.

[155] Pi Y, Li X, Xia Q, et al. Adsorptive and photocatalytic removal of persistent organic pollutants (POPs) in water by metal-organic frameworks (MOFs)[J]. Chemical Engineering Journal, 2017, 337: 351-371.

[156] Sun D, Ye L, Li Z. Visible-light-assisted aerobic photocatalytic oxidation of amines to imines overNH_2-MIL-125(Ti)[J]. Applied Catalysis B Environmental, 2015, 164: 428-432.

[157] Wen M, Li G, Liu H, et al. Metal-organic framework-based nanomaterials for adsorption and photocatalytic degradation of gaseous pollutants: Recent progressand challenges[J]. Environmental Science: Nano, 2019, 6(4): 1006-1025.

[158] Wen M, Mori K, Kuwahara Y, et al. Design of single-site photocatalysts by using metal-organic frameworks as a matrix[J]. Chemistry-An Asian Journal, 2018, 13(14): 1767-1779.

[159] Zhang Z, Li X, Liu B, et al. Hexagonal microspindle of NH_2-MIL-101(Fe) metal-organic frameworks with visible-light-induced photocatalytic activity for the degradation of toluene[J]. Rsc Advances, 2015, 6(6): 4289-4295.

[160] Gong J, Li C, Wasielewski M. Advances in solar energy conversion[J]. Chem Soc Rev, 2019, 48(7): 1862-1864.

[161] Li Z, XiaoJ, Jiang H. Encapsulating a Co(Ⅱ) molecular photocatalyst in metal-organic framework for visible-fight-driven H_2 production: Boosting catalytic efficiency via spatial charge separation[J]. ACS Catal, 2016, 6(8): 5359-5365.

[162] Ziebel M, Gaggioli C, Turkiewicz A, et al. Effects of covalency on anionic redox chemistry in semiquinoid-based metal-organic frameworks[J]. J Am Chem Soc, 2020, 142(5): 2653-2664.

[163] Yoon J, Kim J, Kim C, et al. MOF-based hybrids for solar fuel production[J]. Adv Energy Mater, 2021, 11(27): 2003052.

[164] Hu Y, Zhang L. Amorphization of metal-organic framework MOF-5 at unusually low applied pressure[J]. Phys Rev B, 2010, 81(17): 174103.

[165] Furukawa S, Reboul J, Diring S, et al. Structuring of metal-organic frameworks at the mesoscopic/macroscopic scale[J]. Chem Soc Rev, 2014, 43(16): 5700-5734.

[166] Ding M, Flaig R, Jiang H, et al. Carbon capture and conversion using metal-organic frameworks and MOF-based materials[J]. Chem Soc Rev, 2019, 48(10): 2783-2828.

[167] Fu Y, Sun D, Chen Y, et al. An amine-functionalized titanium metal-organic framework photocatalyst with visible-light-induced activity for CO_2 reduction[J]. Angew Chem Int Ed,

2012, 51(14): 3364-3367.

[168] Wang D, Huang R, Liu W, et al. Fe-based MOFs for photocatalytic CO_2 reduction: Role of coordination unsaturated sites and dual excitation pathways[J]. ACS Catal, 2014, 4(12): 4254-4260.

[169] Zhang H, Hong Q, Li J, et al. Isolated squar-planar copper center in boron imidazolate nanocages for photocatalytic reduction of CO_2 to CO[J]. Angew Chem Int Ed, 2019, 58(34): 11752-11756.

[170] Qiu J, Zhang X, Feng Y, et al. Modified metal-organic frameworks as photocatalysts[J]. Applied Catalysis B-Environmental, 2018, 231: 317-342.

[171] Silva C, Luz, I, Francesc X, et al. Water stable Zr-benzenedicarboxylate metal-organic frameworks as photocatalysts for hydrogen generation[J]. Chemistry-AEuropean Journal, 2010, 16(36): 11133-11138.

[172] Dan-Hardi M, Serre C, Frot T, et al. A new photoactive crystalline highly porous titanium(Ⅳ) dicarboxylate[J]. Journal of the American Chemical Society, 2010, 131(31): 10857-10859.

[173] Fu Y, Sun D, Chen Y, et al. All amine-functionalized titanium metal-organic framework photocatalyst with visible-light-indnced activity for CO_2 reduction[J]. Angewandte Chemie, 2012, 51(14): 3364-3367.

[174] Hendon C, Tiana D, Fontecave M, et al. Engineering the optical response of the Titanium-MIL-125 metal-organic framework through ligand functionalization[J]. Journal of the American Chemical Society, 2013, 135(30): 10942-10945.

[175] Nasalevich M, Goesten M, Savenije T, et al. Enhancing optical absorption of metal-organic frameworks for improved visible light photocatalysis[J]. Chemical Communications, 2013, 49(90): 10575-10577.

[176] Wang C, Xie Z, Krafft K, et al. Doping metal-organic frameworks for water oxidation, carbon dioxide reduction, and organic photocatalysis[J]. J Am Chem Soc, 2011, 133(34): 13445-13454.

[177] Kajiwara T, Fujii M, Tsujimoto M, et al. Photochemical reduction of low concentrations of CO_2 in a porous coordination polymer with a ruthenium(Ⅱ)- COcomplex[J]. Angew Chem Int Ed, 2016, 55(8): 2697-2700.

[178] Wang J, Wu J, Lu L, et al. A new 3D 10-connected Cd(Ⅱ) based MOF with mixed ligands: A dual photoluminescent sensor for nitroaroamatics and ferric ion[J]. Frontiers in Chemistry, 2019, 7: 111-115.

[179] Amarajothi D, Malik A, Garcia H. Mixed-metal or mixed-linker metal organic frameworks as heterogeneous catalysts[J]. Catalysis Science & Technology, 2016, 6(14): 5238-5261.

[180] Min K, Cahill J, Fei H, et al. Postsynthetic ligand and cation exchange in robust metal-organic frameworks[J]. Journal of the AmericanChemical Society, 2012, 134(43): 18082-18088.

[181] Lee Y, Kim S, Fei H, et al. Photocatalytic CO_2 reduction using visible light by metal-monocatecholato species in a metal-organic framework[J]. Chem Commun, 2015, 51(92):

16549-16552.

[182] Wang S, Lin J, Wang X, et al. Semiconductor-redox catalysis promoted by metal-organic frameworks for CO_2 reduction[J]. Phys Chem Chem Phys, 2014, 16(28): 14656.

[183] Wang S, Wang X. Photocatalytic CO_2 reduction by CdS promoted with a zeolitic imidazolate framework[J]. Appl Catal B-Environ, 2015, 162: 494-500.

[184] Liu Q, Low Z, Li L, et al. ZIF-8/Zn_2GeO_4 nanorods with an enhanced CO_2 adsorption property in an aqueous medium for photocatalytic synthesis of liquid fuel[J]. J Mate Chem A, 2013, 1(38): 11563.

[185] Crake A, Christoforidis K, Kafizas A, et al. CO_2 capture and photocatalytic reduction using bifunctional TiO_2/MOF nanocomposites under UV-vis irradiation[J]. Appl Catal B-Environ, 2017, 210: 131-140.

[186] Wu L, Mu Y, Guo X, et al. Encapsulating perovskite quantum dots in iron-based metal-organic frameworks (MOFs) for efficient photocatalytic CO_2 reduction[J]. Angew Chem Int Ed, 2019, 58 (28): 9491-9495.

[187] Xiao J, Han L, Luo J, et al. Back cover: Integration of plasmonic effects and schottky junctions into metal-organic framework composites: Steering charge flow for enhanced visible-light photocatalysis[J]. Angewandte Chemie International Edition, 2018, 57(4): 1103-1107.

[188] Li D, Yu S, Jiang H. From UV to near-infrared light-responsive metal-organic framework composites: Plasmon and upconversion enhanced photocatalysis[J]. Advanced Materials, 2018, 30(27): 1707377.

[189] Hu S, Sun Y, Feng Z, et al. Design and construction strategies to improve covalent organic frameworks photocatalyst's performance for degradation of organic pollutants[J]. Chemosphere, 2021, 286(1): 131646.

[190] Yan S, Ouyang S, Xu H, et al. Co-ZIF-9/TiO_2 nanostructure for superior CO_2 photoreduction activity[J]. J Mate Chem A, 2016, 4 (39): 15126-15133.

[191] YangR, Cai J, LvK, et al. Fabrication of TiO_2 hollow microspheres assembly from nanosheets (TiO_2-HMSs-NSs) with enhanced photoelectric conversion efficiency in DSSCs and photocatalytic activity[J]. Applied Catalysis B: Environmental, 2017, 210: 184-193.

[192] Bie C, Fu J, Cheng B, et al. Ultrathin CdS nanosheets with tunable thickness and efficient photocatalytic hydrogen generation[J]. Applied Surface Science, 2018, 462: 606-614.

[193] Zhao X, Du Y, Zhang C, et al. Enhanced visible photocatalytic activity of TiO_2 hollow boxes modified by methionine for RhB degradation and NO oxidation[J]. Chinese Journal of Catalysis, 2018, 39(4): 736-746.

[194] Zhou M, Wang S, Yang P, et al. Boron carbon nitride semiconductors decorated with CdS nanoparticles for photocatalytic reduction of CO_2[J]. ACS Catalysis, 2018, 8(6): 4928-4936.

[195] Li K, Zhang S, Li Y, et al. MXenes as noble-metal-alternative co-catalysts in photocatalysis[J]. Chinese Journal of Catalysis, 2021, 42(1): 3-14.

[196] Qian R, Zong H, Schneider J, et al. Charge carrier trapping, recombination and transfer during TiO_2 photocatalysis: An overview[J]. Catalysis Today, 2019, 335: 78-90.

[197] Negrin-Montecelo Y, Testa-Anta M, Marin-Caba L, et al. Titanate nanowires as one-dimensional hot spot generators for broadband Au-TiO_2 photocatalysis[J]. Nanomaterials, 2019, 9(7): 990.

[198] Sohn Y, Huang W, Taghipour F. Recent progress and perspectives in the photocatalytic CO_2 reduction of Ti-oxide-based nanomaterials[J]. Applied Surface Science, 2017, 396: 1696-1711.

[199] Joo J, Zhang Q, Dahl M, et al. Control of the nanoscale crystallinity in mesoporous TiO_2 shells for enhanced photocatalytic activity[J]. Energy & Environmental Science, 2012, 5(4): 6321-6327.

[200] Zhang H, Liu L, Zhou Z. Towards better photocatalysts: First-principles studies of the alloying effects on the photocatalytic activities of bismuth oxyhalides under visible light[J]. Physical Chemistry Chemical Physics, 2012, 14(3): 1286-1292.

[201] An H, Du Y, Wang T, et al. Photocatalytic properties of BiOX (X=Cl, Br, and I)[J]. Rare Metals, 2008, 27(3): 243-250.

[202] Li J, Yu Y, Zhang L. Bismuth oxyhalide nanomaterials: Layered structures meet photocatalysis[J]. Nanoscale, 2014, 6(15): 8473-8488.

[203] Di J, Xia J, Li H, et al. Bismuth oxyhalide layered materials for energy and environmental applications[J]. Nano Energy, 2017, 41: 172-192.

[204] Di J, Xia J, Ge Y, et al. Reactable ionic liquid-assisted rapid synthesis of BiOI hollow microspheres at room temperature with enhanced photocatalytic activity[J]. Journal of Materials Chemistry A, 2014, 2(38): 15864-15874.

[205] Zhao L, Zhang X, Fan C, et al. First-principles study on the structural, electronic and optical properties of BiOX (X=Cl, Br, I) crystals[J]. Physica B, 2012, 407(17): 3364-3370.

[206] Chen L, Huang R, Xiong M, et al. Room-temperature synthesis of flower-like BiOX (X = Cl, Br, I) hierarchical structures and their visible-light photocatalytic activity[J]. Inorganic Chemistry, 2013, 52 (19): 11118-11125.

[207] Tian N, Huang H, Wang S, et al. Facet-charge-induced coupling dependent interfacial photocharge separation: A case of BiOI/g-C_3N_4 p-n junction[J]. Applied Catalysis B: Environmental, 2020, 267: 118697.

[208] Huo Y, Zhang J, Miao M, et al. Solvothermal synthesis of flower-like BiOBr microspheres with highly visible-light photocatalytic performances[J]. Applied Catalysis B: Environmental, 2012, 111-112: 334-341.

[209] Drache M, Roussel P, Wignacourt J. Structures and oxide mobility in Bi-Ln-O materials: Heritage of Bi_2O_3[J]. Chemical Reviews, 2007, 107(1): 80-96.

[210] Xiao X, Hu R, Liu C, et al. Facile large-scale synthesis of β-Bi_2O_3 nanospheres as a highly efficient photocatalyst for the degradation of acetaminophen under visible light irradiation[J]. Applied Catalysis B: Environmental, 2013, 140: 433-443.

[211] Wang W, Chen X, Liu G, et al. Monoclinic dibismuth tetraoxide: A new visible-light-driven photocatalyst for environmental remediation[J]. Applied Catalysis B: Environmental, 2015, 176-177: 444-453.

[212] Eberl J, Kisch H. Visible light photo-oxidations in the presence of α-Bi_2O_3[J]. Photochemical & Photobiological Sciences, 2008, 7 (11): 1400-1406.

[213] Zhang G, Chen D, Li N, et al. Fabrication of Bi_2MoO_6/ZnO hierarchical heterostructures with enhanced visible-light photocatalytic activity[J]. Appl Catal B Environ, 2019, 250: 313-324.

[214] Wang J, Tang L, Zeng G, et al. 0D/2D interface engineering of carbon quantum dots modified Bi_2WO_6 ultrathin nanosheets with enhanced photoactivity for full spectrum light utilization and mechanism insight[J]. Appl Catal B Environ, 2018, 222: 115-123.

[215] Fu H, Pan C, Yao W, et al. Visible-light-induced degradation of Rhodamine B by nanosized Bi_2WO_6[J]. J Phys Chem B, 2005, 109(47): 22432-22439.

[216] Tian H, Teng F, Xu J, et al. An innovative anion regulation strategy for energy bands of semiconductors: A case from Bi_2O_3 to $Bi_2O(OH)_2SO_4$[J]. Sci Rep-UK, 2015, 5: 1-9.

[217] Kudo A, Hijii S. H_2 or O_2 evolution from aqueous solutions on layered oxide photocatalysts consisting of Bi^{3+} with 6s(2) configuration and d(0) transition metal ions[J]. Chemistry Letters, 1999, (10): 1103-1104.

[218] Tang J, Zou Z, Ye J. Photocatalytic decomposition of organic contaminants by Bi_2WO_6 under visible light irradiation[J]. Catalysis Letters, 2004, 92(1-2): 53-56.

[219] Zhang N, Ciriminna R, Pagliaro M, et al. Nanochemistry-derived Bi_2WO_6 nanostructures: Towards production of sustainable chemicals and fuels induced by visible light[J]. Chemical Society Reviews, 2014, 43(15): 5276-5287.

[220] Zhang L, Wang H, Chen Z, et al. Bi_2WO_6 Micro/nano-structures: Synthesis, modifications and visible-light-driven photocatalytic applications[J]. Applied Catalysis B: Environmental, 2011, 106(1-2): 1-13.

[221] Zhang L, Wang W, Chen Z, et al. Fabrication of flower-like Bi_2WO_6 superstructures as high performance visible-light driven photocatalysts[J]. Journal of Materials Chemistry, 2007, 17(24): 2526-2532.

[222] Zhang C, Chen G, Lv C, et al. Enabling nitrogen fixation on Bi_2WO_6 photocatalyst by c-PAN surface decoration[J]. ACS Sustain Chem Eng, 2018, 6: 11190-11195.

[223] Shimodaira Y, Kato H, Kobayashi H, et al. Photophysical properties and photocatalytic activities of bismuth molybdates under visible light irradiation[J]. The Journal of Physical Chemistry B, 2006, 110(36): 17790-17797.

[224] Di J, Xia J, Ji M, et al. The synergistic role of carbon quantum dots for the improved photocatalytic performance of Bi_2MoO_6[J]. Nanoscale, 2015, 7: 11433-11443.

[225] Alarcon-Llado E, Chen L, Hettick M, et al. $BiVO_4$ thin film photoanodes grown by chemical vapor deposition[J]. Physical Chemistry Chemical Physics, 2014, 16: 1651.

[226] Zhao Y, Li R, Mu L, et al. The significance of crystal morphology controlling in semiconductor-based photocatalysis: A case study on $BiVO_4$ photocatalyst[J]. Crystal Growth & Design, 2017, 17(6): 2923-2928.

[227] Tan H, Amal R, Ng Y. Alternative strategies in improving the photocatalytic and photoelectrochemical activities of visible light-driven $BiVO_4$: A review [J]. Journal of Materials Chemistry A, 2017, 5(32): 16498-16521.

[228] Park Y, McDonald K J, Choi K S. Progress in bismuth vanadate photoanodes for use in solar water oxidation[J]. Chemical Society Reviews, 2013, 42(6): 2321-2337.

[229] Cooper J, Gul S, Toma F, et al. Electronic structure of monoclinic $BiVO_4$ [J]. Chemistry of Materials, 2014, 26(18): 5365-5373.

[230] Xi G, Ye J. Synthesis of bismuth vanadate nanoplates with exposed {001} facets and enhanced visible-light photocatalytic properties [J]. Chemical Communications, 2010, 46(11): 1893-1895.

[231] Sun Y, Xie Y, Wu C, et al. Aqueous synthesis of mesostructured $BiVO_4$ quantum tubes with excellent dual response to visible light and temperature[J]. Nano Research, 2010, 3(9): 620-631.

[232] Zhou J, Tian G, Chen Y, et al. Growth rate controlled synthesis of hierarchical Bi_2S_3/In_2S_3 core/shell microspheres with enhanced photocatalytic activity[J]. Sci Rep, 2014, 4: 4027.

[233] Low J, Yu J, Jaroniec M, et al. Heterojunction photocatalysts[J]. Advanced Materials, 2017, 29(20): 1601694.

[234] Lu H, Hao Q, Chen T, et al. A high-performance Bi_2O_3/Bi_2SiO_5 p-n heterojunction photocatalyst induced by phase transition of Bi_2O_3 [J]. Applied Catalysis B: Environmental, 2018, 237: 59-67.

[235] Han C, Wu L, Ge L, et al. AuPd bimetallic nanoparticles decorated graphitic carbon nitride for highly efficient reduction of water to H_2 under visible light irradiation[J]. Carbon, 2015, 92(18): 31-40.

[236] Rayalu S, Jose D, Mangrulkar P, et al. Photodeposition of AuNPs on metal oxides: Study of SPR effect and photocatalytic activity[J]. Int J Hydrogen Energ, 2014, 39(8): 3617-3624.

[237] Chen Y, Huang W, He D, et al. Construction of heterostructured g-C_3N_4/Ag/TiO_2 microspheres with enhanced photocatalysis performance under visible-light irradiation[J]. ACS Appl Mater Inter, 2014, 6(16): 14405-14414.

[238] Jiang S, Xiong C, Song S, et al. Plasmonic graphene-like Au/C_3N_4 nanosheets with barrier-free interface for photocatalytically sustainable evolution of active oxygen species[J]. ACS Sustain Chem Eng, 2019, 7(2): 2018-2026.

[239] Luo B, Xu D, Li D, et al. Fabrication of a Ag/Bi_3TaO_7 plasmonic photocatalyst with enhanced photocatalytic activity for degradation of tetracycline[J]. ACS Appl Mater Interfaces, 2015, 7(31): 17061.

[240] Hayashido Y, Naya S, Tada H, et al. Local electric field-enhanced plasmonic photocatalyst:

Formation of Ag cluster-incorporated AgBr nanoparticles on TiO$_2$[J]. J Phys Chem C, 2016, 120(35): 19663-19669.

[241] Zhang M, Shao C, Li X, et al. Carbon-modified BiVO$_4$ microtubes embedded with Ag nanoparticles have high photocatalytic activity under visible light[J]. Nanoscale, 2012, 4(23): 7501-7508.

[242] Niu F, Chen D, Qin L, et al. Synthesis of Pt/BiFeO$_3$ heterostructured photocatalysts for highly efficient visible-light photocatalytic performances[J]. Sol Energ Mat Sol C, 2015, 143: 386-396.

[243] Dong F, Xiong T, Sun Y, et al. A semimetal bismuth element as a direct plasmonic photocatalyst[J].Commun, 2014, 50: 10386-10394.

[244] Wang T, Jin B, Jiao Z, et al.Photo-directed growth of Au nanowires on ZnO arrays for enhancing photoelectrochemical performances[J]. J Mater Chem A, 2014, 2(37): 15553-15561.

[245] Jiao Z, Shang M, Liu J, et al. The charge transfer mechanism of Bi modified TiO$_2$ nanotube arrays: TiO$_2$ serving as a "charge-transfer-bridge"[J]. Nano Energy, 2017, 31: 96.

[246] Wang H, Zhang L, Chen Z, et al. Semiconductor heterojunction photocatalysts: Design, construction, and photocatalytic performances[J]. Chem Soc Rev, 2014, 43(15): 5234-5244.

[247] Marschall R. Semiconductor composites: Strategies for enhancing charge carrier separation to improve photocatalytic activity[J]. Adv Funct Mater, 2014, 24(17): 2421-2440.

[248] Moniz S, Shevlin S, Martin D, et al. Visible-light driven heterojunction photocatalysts for water splitting-a critical review[J]. Energ Environ Sci, 2015, 8(3): 731-759.

[249] Wei N, Liu Y, Feng M, et al. Controllable TiO$_2$ core-shell phase heterojunction for efficient photoelectrochemical water splitting under solar light[J]. Appl Catal B Environ, 2019, 244: 519-528.

[250] Zhang Z, Huang J, Zhang M, et al. Ultrathin hexagonal SnS$_2$ nanosheets coupled with g-C$_3$N$_4$ nanosheets as2D/2D heterojunction photocatalysts toward high photocatalytic activity[J]. Appl Catal B Environ, 2015, 163: 298-305.

[251] Li W, Tian Y, Li H, et al. Novel BiOCl/TiO$_2$ hierarchical composites: Synthesis, characterization and application on photocatalysis[J]. Appl Catal A Gen, 2016, 516: 81-89.

[252] Wang K, Zhang Y, Lu N. Materials BiOBr nanosheets-decorated TiO$_2$ nanofibers as hierarchical p-n heterojunctions photocatalysts for pollutant degradation[J]. J Mater Sci, 2019, 54: 8426-8435.

[253] Zhou P, Yu J, Jaroniec M. All-solid-state z-scheme photocatalytic systems[J]. Adv Mater, 2014, 26(29): 4920-4935.

[254] Tian L, Xian X, Cui X, et al. Fabrication of modified g-C$_3$N$_4$ nanorod/Ag$_3$PO$_4$ nanocomposites for solar-driven photocatalytic oxygen evolution from water splitting[J]. Appl Surf Sci, 2018, 430: 301-308.

[255] Wang Q, Hisatomi T, Jia Q, et al. Scalable water splitting on particulate photocatalyst sheets with a solar-to-hydrogen energy conversion efficiency exceeding 1%[J]. Nat Mater, 2016, 15(6): 611-615.

[256] Ma T, Wu J, Mi Y, et al. Novel Z-scheme g-C_3N_4/C@Bi_2MoO_6 composite with enhanced visible-light photocatalytic activity for B-naphthol degradation[J]. Sep Purif Technol, 2017, 183: 54-65.

[257] Wan J, Xue P, Wang R, et al. Synergistic effects in simultaneous photocatalytic removal of Cr (Ⅵ) and tetracycline hydrochloride by Z-scheme Co_3O_4/Ag/Bi_2WO_6 heterojunction[J]. Appl Surf Sci, 2019, 483: 677-687.

[258] Hao J, Zhang S, Ren F, et al. Synthesis of TiO_2@g-C_3N_4 core-shell nanorod arrays with Z-scheme enhanced photocatalytic activity under visible light[J]. J Colloid Interf Sci, 2017, 508: 419-425.

[259] Zhang Z, Huang J, Fang Y, et al. A Nonmetal plasmonic Z-scheme photocatalyst with UV- to NIR-driven photocatalytic protons reduction[J]. Adv Mater, 2017, 29(18): 1606688.

[260] Wang X, Zhou X, Shao C, et al. Graphitic carbon nitride/BiOI loaded on electrospun silica nanofibers with enhanced photocatalytic activity[J]. Appl Surf Sci, 2018, 455: 952-962.

[261] Chen X, Li Y, Li L. Facet-engineered surface and interface design of WO_3/Bi_2WO_6 photo- catalyst with direct Z-scheme heterojunction for efficient salicylic acid removal[J]. Appl Surf Sci, 2020, 508: 1-10.

[262] Wu R, Zhou K, Yue C, et al. Recent progress in synthesis, properties andpotential applications of SiC nanomaterials[J]. Progress in Materials Science, 2015, 72: 1-60.

[263] Beberwyck B, Surendranath Y, Alivisatos A. Cation exchange: A versatiletool for nanoma- terials synthesis[J]. The Journal of Physical Chemistry C, 2013, 117(39): 19759-19770.

[264] Wang H, Rogach A. Hierarchical SnO_2 nanostructures: Recent advances indesign, synthesis, and applications[J]. Chemistry of Materials, 2014, 26(1): 123-133.

[265] Yang S, Shao C, Tao R, et al. Enhanced full-spectrum-response photocatalysis and reusa- bility of $MoSe_2$ via hierarchical N-doped carbon nanofibers as heterostructural supports[J]. ACS Sustain Chem Eng, 2018, 6(11): 14314-14322.

[266] Zhang M, Shao C, Guo Z, et al. Highly efficient decomposition of organic dye by aqueous- solid phase transfer and in situ photocatalysis using hierarchical copper phthalocyanine hollow spheres[J]. ACS Appl Mater Inter, 2011, 3: 2573-2578.

[267] Rezaee S, Shahrokhian S. Facile synthesis of petal-like NiCo/NiO-CoO/nanoporous carbon composite based on mixed-metallic MOFs and their application for electrocatalytic oxidation of methanol[J]. Appl Catal B Environ, 2019, 244: 802-813.

[268] Gang Liu, Peng Cui, Liu X, et al. A facile preparation strategy for Bi_2O_4/Bi_2WO_6 heterojun- ction with excellent visible lightphotocatalytic activity[J]. Journal of Solid State Chemistry 2020, 290: 121542-121550.

[269] Darrell H, Alexander L, Hao F, et al. Bending instability of electrically charged liquid jets of polymer solutions in electrospinning[J]. J Appl Phys, 2000, 87(9): 4531-4547.

[270] Xie J, Li X, Xia Y. Putting electrospun nanofibers to work for biomedical research[J]. Macromol Rapid Comm, 2008, 29(22): 1775-1792.

[271] Greiner A, Wendorff J. Electrospinning: A fascinating method for the preparation of ultrathin fibers[J]. Angew Chem Int Ed, 2007, 46(30): 5670-5703.

[272] Guo X, Zhou X, Li X, et al. Bismuth oxychloride (BiOCl)/copper phthalocyanine (CuTNPc) heterostructures immobilized on electrospun polyacrylonitrile nanofibers with enhanced activity for floating photocatalysis[J]. J Colloid Interf Sci, 2018, 525: 187-195.

第二章

Bi_2WO_6/MIL-100(Fe)复合光催化材料制备及性能研究

2.1 引言   83
2.2 实验部分   84
2.3 结果与讨论   86
2.4 结论   92
    参考文献   92

2.1 引言

半导体光催化剂因其在利用丰富的太阳能解决全球环境和能源相关问题方面具有巨大潜力而备受关注[1-5]。对各种光催化剂，例如 TiO_6[6]、ZnO[7]，在水分解、污染物降解或太阳能电池方面进行了研究。由于半导体光催化剂的宽带隙（>3eV），它们中的大多数只能吸收紫外光，这极大地限制了它们的实际应用。铋基氧化物光催化剂其禁带宽度一般分布在 2~3eV 之间，能被波长位于 420~600nm 的光所激发，而此波段覆盖了大部分可见光，因此在光催化领域有着巨大的应用潜力，如钙钛矿型 Bi_2WO_6 带隙为 2.75eV、单斜晶系的 $BiVO_4$ 带隙为 2.3~2.4eV、四方晶系的 $BiOI$[11] 禁带宽度为 1.77eV 等[8-13]。铋基氧化物光催化剂虽然具有很强的可见光响应能力，但其自身载流子复合率高、光化学反应活性低，限制了其光催化应用，这已成为铋基氧化物等可见光光催化材料目前迫切需要解决的关键问题。为了改善其光催化活性，科研工作者已经做出了许多努力，例如改变形貌[14]、提高结晶度[15]、半导体复合[16]和离子掺杂[17]。研究表明，光催化剂与其他材料复合，可促进光生载流子的有效分离，可以提高其光催化效率。如制备的 Bi_2WO_6 和 TiO_2、Ag_3PO_4、$BiOCl$ 以及 ZnO 异质结构光催化剂，明显提高了载流子的分离效率，同时提高了其可见光光催化活性[18-21]，但是文献所报道的 Bi_2WO_6 材料大部分为粉末态，颗粒之间易团聚，而且粉末材料在溶液反应中的分离非常困难，因此，设计新型高比表面积、易回收的 Bi_2WO_6 复合光催化材料是非常必要的。与这些传统的无机材料相比，金属有机骨架（金属有机骨架化合物）具有较大的孔隙率和特定的表面，这使其成为光催化应用中与半导体复合的候选材料。但是据我们所知，目前尚无关于钨酸铋/金属有机骨架化合物异质结制备的报道。金属有机骨架化合物是一类新型的晶体多孔材料，由金属离子和有机配体构成[22-24]，由于其超高的比表面积、完美的框架结构、可调节功能以及各种潜在的应用，引起了科研工作者极大的兴趣[25-27]。研究表明光辐照金属有机骨架化合物会导致电子和空穴的产生[28-33]，因此，将金属有机骨架化合物用作光催化剂在理论上是可行的[34]。然而，由于单组分金属有机骨架化合物的电子电导率较差，因此其光催化应用效率低下[23-28]，因此，迫切需要解决此问题，将金属有机骨架化合物与其他高导电性无机半导体相结合可能是一种理想的方法[35-36]。

在已报道的一系列金属有机骨架材料中，过渡金属与均苯三甲酸构成的 MIL-100 具有比表面积大、孔体积大、热稳定性和化学稳定性高等优点，近年来受到科研工作者的广泛关注。相比其他由铜、钴、铬等金属构成的材料，MIL-100（Fe）具有无毒无害、价格低廉、对环境无污染且铁具有独特的氧化还原性能、易于工业化推广等诸多优点。MIL-100（Fe）骨架结构中有两种介孔笼，孔直径分别为 2.5nm 和 2.9nm，具有较高的比表面积，热稳定温度在 270℃左右。由于该骨架材料含有铁，在不破坏骨架结构的条件下，三价铁可在一定条件下活化，铁-氧（Fe-O）团簇通过芬顿效应将过氧化氢活化为羟基自由基（·OH），因此 Fe 基金属有机骨架化合物在可见光区域表现出强烈的吸收作用[37-39]。因此在这项工作中，我们采用溶剂热法将 MIL-100（Fe）与 Bi_2WO_6 复合制备新型的 p-n 结光催化材料，合成了一系列具有不同 MIL-100（Fe）含量的 Bi_2WO_6/MIL-100（Fe）纳米复合材料。研究结果表明，所有的 Bi_2WO_6/MIL-100（Fe）异质结构都比纯 Bi_2WO_6 和 MIL-100（Fe）表现出更好的光催化活性，并且当 MIL-100（Fe）的质量分数为 3.5% 时，光催化活性最好。这可以归因于 Bi_2WO_6 和 MIL-100（Fe）形成的 p-n 结促进了光生载流子的有效分离（图 2-1）。循环实验表明，Bi_2WO_6/MIL-100（Fe）纳米复合材料具有较高的光催化稳定性。

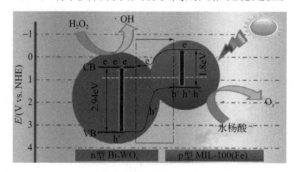

图 2-1　Bi_2WO_6/MIL-100（Fe）p-n 结光生载流子转移机制

2.2　实验部分

2.2.1　实验方案

2.2.1.1　MIL-100（Fe）的制备。

MIL-l00（Fe）纳米颗粒是依据文献制备的[19]，分别称取 0.8456g 均苯三甲酸和 1.6253g $FeCl_3 \cdot 6H_2O$ 溶解于 30mL 蒸馏水中，然后加入 0.214mL HF 和 0.165mL HNO_3，搅拌使其均匀混合后，将混合好的溶液转移到 50mL 的不锈钢反应釜中，加热到 150℃，然后反应 12h，待反应结束后，通过离心分离方法回收所制备的橘黄色产物，用热的蒸馏水洗涤 3 次，最后于 150℃真空干燥 12h。

2.2.1.2 Bi_2WO_6 的制备。

准确称取 0.2mmol $Bi(NO_3)_3 \cdot 5H_2O$ 溶解于 15mL 的乙二醇中并超声 30min，同时称取 0.1mmol $Na_2WO_4 \cdot 2H_2O$ 溶解于 5mL 的乙二醇中并超声 30min。将溶解好的两种物质混合，并将混合液超声 30min 使其混合均匀。在混合均匀的溶液中加入 10mL 无水乙醇，并混合均匀。将混合均匀后的溶液转移到 50mL 的不锈钢反应釜中，180℃下反应 36h。通过离心分离的手段回收制得的产物，用蒸馏水洗涤 3 次，最后 85℃干燥 12h。

2.2.1.3 Bi_2WO_6/MIL-100（Fe）复合材料的制备。

称取上面制备好的 MIL-100（Fe），加入溶解有 Na_2WO_4 的 5mL 乙二醇溶液中，超声 30min，使 Na_2WO_4 分散在 MIL-100（Fe）孔洞之中。加入 30mL $Bi(NO_3)_3$ 的乙二醇溶液，超声 30min，使其与带有 MIL-100（Fe）的 Na_2WO_4 溶液混合均匀，超声 10min。将混合好的溶液转入 50mL 的不锈钢反应釜之中，160℃下反应 24h。

2.2.2 表征

在具有 CuKα X 射线衍射（λ=0.15406nm）的布鲁克 AXS D8 Advance 粉末衍射仪上表征了样品的 X 射线粉末衍射（XRD）。使用日本电子（JEOL）Hitachi S-4800 场发射扫描电子显微镜（FESEM）表征合成样品的形貌。用 JEOL JEM-2100 显微镜获得了高分辨率的透射电子显微镜（HRTEM）图像。样品的紫外 - 可见漫反射光谱（DRS）在紫外 - 可见光谱仪（日本岛津，UV-2501PC）上记录，积分球附着在 200～2000nm 范围内，并以硫酸钡为参考。使用比表面分析仪（美国麦克仪器，ASAP2020），使用 Brunauer-Emmett-Teller（BET）方法计算比表面积。

2.2.3 光催化活性测试

在实验装置中，使用 300W 氙灯作为光源，并使用 420nm 截止滤光片并仅提供可见光照射。将约 100mg 的光催化剂添加到 100mL 的水杨酸溶液中

（c=10mg/L）。在光激化之前，将悬浮液在黑暗中磁力搅拌3h，以达到光催化剂与水杨酸之间的吸附-解吸平衡。然后在磁力搅拌下使溶液暴露于可见光照射范围内。从悬浮液中收集约3mL的等分试样，每过10min立即离心。通过使用紫外-可见光谱仪检测在296nm处的吸光度来监测水杨酸的降解。水杨酸的浓度通过其校准曲线计算。降解比为c_t/c_0。c_t是在吸收光谱主峰的每个照射时间间隔内水杨酸的浓度。c_0是达到吸附-解吸平衡时的初始浓度。

实验中的主要步骤及技术路线如图2-2所示

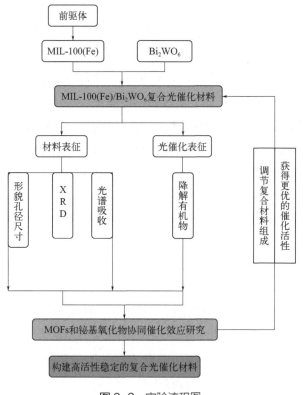

图2-2　实验流程图

2.3　结果与讨论

通过简单的水热共沉淀法合成了一系列不同MIL-100（Fe）含量的Bi_2WO_6/MIL-100（Fe）复合光催化材料。首先MIL-100（Fe）是通过典型的水热法制备的，然后在Bi_2WO_6前驱体溶液中添加不同量的MIL-100（Fe）。

分别将 30、60、90 和 120mg 的 MIL-100（Fe）添加到溶液中，并将样品命名为 Bi_2WO_6/MIL-100（Fe）-1、Bi_2WO_6/MIL-100（Fe）-2、Bi_2WO_6/MIL-100（Fe）-3、Bi_2WO_6/MIL-100（Fe）-4。合成产物的 X 射线衍射图谱如图 2-3 所示。纯 Bi_2WO_6 和 MIL-100（Fe）以及模拟的 MIL-100（Fe）的 XRD 也在图 2-3 中。XRD 图谱中 Bi_2WO_6 的特征衍射峰与斜方晶 Bi_2WO_6（JCPDS 号 39-0256）[40]一致。XRD 图谱中 MIL-100（Fe）的特征衍射峰与立方晶体 MIL-100（Fe）[41]一致。在 Bi_2WO_6/MIL-100（Fe）复合光催化材料体系中不仅能够看到 Bi_2WO_6 的特征衍射峰，而且可以看见部分 MIL-100（Fe）特征衍射峰，随着 MIL-100（Fe）含量的增加特征衍射峰的强度变大，表明成功制备了 Bi_2WO_6/MIL-100（Fe）复合光催化材料。另外在 Bi_2WO_6/MIL-100（Fe）复合光催化材料 XRD 图谱中没有观察到杂质峰，这表明所制备的样品纯度比较高，而且 MIL-100（Fe）在水热处理过程中保持其晶格结构完整，MIL-100（Fe）的存在并不妨碍 Bi_2WO_6 晶体的形成。

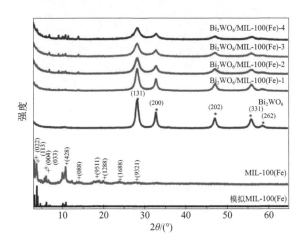

图 2-3 各样品的 XRD 图谱 [图底部是根据 MIL-100（Fe）的晶体学数据模拟的 XRD 图谱]

为分析 Bi_2WO_6/MIL-100（Fe）复合光催化材料的形貌，本文分别利用扫描电子显微（SEM）和高分辨率透射电子显微镜（HRTEM）对样品进行分析。如图 2-4 所示，图 2-4（a）为 MIL-100（Fe）的 SEM 图，从图中可以看出 MIL-100（Fe）主要由平均边长为 1μm 的八面体微晶和平均边长为 200nm 的小八面体微晶组成。图 2-4（b）为 Bi_2WO_6 的 SEM 图，如图所示 Bi_2WO_6

是平均尺寸为 20nm 的纳米颗粒。图 2-4（c）为 Bi_2WO_6/MIL-100（Fe）复合光催化材料的 TEM 图像，如图所示，尺寸为 20nm 的 Bi_2WO_6 纳米颗粒附着在 MIL-100（Fe）微晶表面。附着的 Bi_2WO_6 纳米晶体的晶格条纹间距为 0.319nm，对应于 Bi_2WO_6 的（131）晶面［如图 2-4（d）］。但是，无法获得 MIL-100（Fe）的选定区域电子衍射（SAED）和 HRTEM 图像，这可能是因为 MIL-100（Fe）微晶在高能电子束辐照下易于损坏。

图 2-4 （a）MIL-100（Fe）SEM 图像；（b）Bi_2WO_6 SEM 图像；（c）Bi_2WO_6/MIL-100（Fe）TEM 图像；（d）Bi_2WO_6/MIL-100（Fe）HRTEM 图像

表 2-1 MIL-100（Fe）、Bi_2WO_6 和不同质量比的 Bi_2WO_6/MIL-100（Fe）纳米复合材料的多孔结构参数

样品	BET 比表面积/($m^2 \cdot g^{-1}$)	总孔容/($cm^3 \cdot g^{-1}$)
MIL-100（Fe）	1370.0048	0.685811
Bi_2WO_6	86.6988	0.155289
Bi_2WO_6/MIL-100（Fe）-1	99.6509	0.174155
Bi_2WO_6/MIL-100（Fe）-2	139.7424	0.217472
Bi_2WO_6/MIL-100（Fe）-3	168.7056	0.221701
Bi_2WO_6/MIL-100（Fe）-4	213.3622	0.235899

根据样品的 BET 比表面积在 77 K 下的氮吸附-解吸等温线，得到表 2-1 的多孔结构参数如表 2-1 所示，MIL-100（Fe）的含量极大地影响了样品的 BET 比表面积。与纯 Bi_2WO_6 相比，样品的 BET 比表面积随 MIL-100（Fe）

含量的增加而逐渐增加,从 99.6509$m^2 \cdot g^{-1}$ 增至 213.3622$m^2 \cdot g^{-1}$。更大的光催化剂比表面积可以提供更多的表面活性位点并促进电荷载流子的运输,从而促进光催化性能的提高。

制备的光催化材料的光吸收性能用紫外-可见光谱仪进行了测试。如图 2-5 所示,MIL-100(Fe) 和 Bi_2WO_6 的吸收边分别出现在 485nm 和 450nm 左右,与之前报道的结果[40,42]一致。Bi_2WO_6/MIL-100(Fe) 复合光催化材料的纳米颗粒表现出混合吸收特性。相比于纯 Bi_2WO_6 和 MIL-100(Fe),Bi_2WO_6/MIL-100(Fe) 的吸收边随 MIL-100(Fe) 含量的增加而逐渐发生红移。

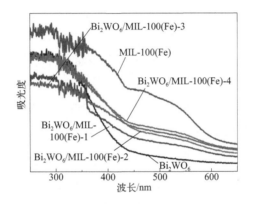

图 2-5 MIL-100(Fe)、Bi_2WO_6 和不同质量比的 Bi_2WO_6/MIL-100(Fe) 纳米复合材料的紫外-可见漫反射光谱

制备的 Bi_2WO_6/MIL-100(Fe) 复合光催化材料的光催化活性通过在可见光的照射下水杨酸的光降解来评估。图 2-6(a) 显示了可见光照射下的不同光催化剂光降解率、水杨酸的浓度与照射时间的关系。没有光催化剂时,水杨酸不会自行光降解,在 50min 内几乎没有观察到变化,表明水杨酸对于入射光照射非常稳定。据悉,存在的少量过氧化氢可以协同产生羟基自由基,从而进一步增强铁基金属有机骨架化合物的光催化活性[40,43]。因此将少量的过氧化氢添加到反应液中用于水杨酸的降解,有可能观察到在没有光催化剂的情况下少量的纯过氧化氢可以将降解率提高到 33.5%,这是因为在可见光照射下形成了羟基自由基:

$$H_2O_2 + 可见光 \longrightarrow \cdot OH + OH^-$$

MIL-100(Fe) 的添加不能进一步改善降解率表明 MIL-100(Fe) 本

身不能用作高效的光催化剂。就 Bi_2WO_6 而言，辐照 50min 后，约 65% 的水杨酸分解。可以看出 Bi_2WO_6/MIL-100（Fe）的异质结构均表现出高于 MIL-100（Fe）和 Bi_2WO_6 的光催化活性。具体而言，Bi_2WO_6/MIL-100（Fe）纳米复合光催化材料的活性随着 MIL-100（Fe）含量从 30mg［Bi_2WO_6/MIL-100（Fe）-1］增加到 60mg［Bi_2WO_6/MIL-100（Fe）-2］而逐渐提高。如果将更多的 MIL-100（Fe）添加到 Bi_2WO_6 中，降解率会再次降低。因此，所制备的 Bi_2WO_6/MIL-100（Fe）-2 纳米复合光催化材料活性是最佳的，辐照 50min 后水杨酸的降解率可达到 95.1%。因此，这些结果清楚地表明，MIL-100（Fe）在增强光催化活性方面起着重要作用。多孔的 MIL-100（Fe）具有较大的比表面积，可以增强水杨酸的吸附。适量的 MIL-100（Fe）可以有效地增大 Bi_2WO_6/MIL-100（Fe）纳米复合材料的比表面积并增强其吸附能力。但是，随着 MIL-100（Fe）的比例进一步增加，会覆盖过多 MIL-100（Fe）而减少 Bi_2WO_6 的含量，因此导致光催化活性迅速降低。因此，只有适当质量分数（3.5%）的 MIL-100（Fe）与 Bi_2WO_6 复合，Bi_2WO_6/MIL-100（Fe）纳米复合材料的光催化效率才能达到最大值。通过重复催化剂对水杨酸的降解来评价异质结光催化剂的稳定性。在可见光下使用 Bi_2WO_6/MIL-100（Fe）-2 作为样品。在每个测试中，光催化剂通过简单过滤后即可重复使用。

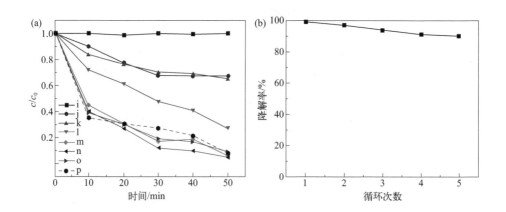

图 2-6 （a）在可见光 i、过氧化氢 j、MIL-100（Fe）k、Bi_2WO_6 l、Bi_2WO_6/MIL-100（Fe）-1 m、Bi_2WO_6/MIL-100（Fe）-2 n、Bi_2WO_6/MIL-100（Fe）-3 o、Bi_2WO_6/MIL-100（Fe）-4 p 下水杨酸降解率；（b）对 Bi_2WO_6/MIL-100（Fe）-2 纳米复合材料进行水杨酸降解的循环测试

如图 2-6（b）所示，光催化在连续五个周期后，在可见光照射下水杨酸的催化降解率并未显示出明显的损失，并且在光催化过程中没有被光腐蚀。表明 Bi_2WO_6/MIL-100（Fe）-2 具有出色的长期稳定性。

因此，我们提出了 Bi_2WO_6/MIL-100（Fe）纳米复合材料光催化降解水杨酸的可能机理（如图 2-7）。MIL-100（Fe）是由有机配体和 Fe-O 原子链构成的三维多孔固体。含有过渡金属作为结构节点的 MIL-100（Fe）也有望成为半导体，因为有机配体的 LUMO 与金属的 d 空轨道混合将形成导带，Zhu 等人以前曾对此进行过报道。MIL-100（Fe）是 p 型半导体[34]，其导带（CB）和价带（VB）的带边位置分别约为 0.45eV 和 1.35eV。此外，Bi_2WO_6 是 n 型半导体，Bi_2WO_6 的 CB 和 VB 的带边位置分别约为 0.39eV 和 3.33eV[44]。因此，与纯 Bi_2WO_6 和 MIL-100（Fe）相比，Bi_2WO_6/MIL-100（Fe）纳米复合材料增强的光催化活性可归因于 Bi_2WO_6 和 MIL-100（Fe）之间的协同作用，p-n 结的形成。图 2-7 显示了 Bi_2WO_6/MIL-100（Fe）异质结构中的电荷转移途径。Bi_2WO_6 的 CB 位于 MIL-100（Fe）的 VB 和 CB 之间，Bi_2WO_6 的 VB 比 MIL-100（Fe）的 VB 低 0.83 eV，因此，Bi_2WO_6/MIL-100（Fe）复合材料属于"B 型异质结"类别[45]。在可见光辐照下，Bi_2WO_6 的 VB 中的电子被激发到 CB 上，MIL-100（Fe）的 CB 上的光生电子将被转移到 Bi_2WO_6 的 CB 上。Bi_2WO_6 的 VB 中的光生空穴将转移到 MIL-100（Fe）的 VB 中。界面处的电场促进电子

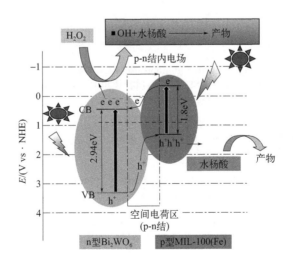

图 2-7　p-n 结的形成和可能的电荷分离过程的简化示意图

从 p 型半导体 CB 迁移到 n 型半导体的 CB。这些将导致 Bi_2WO_6 表面上光生电子聚集和 MIL-100（Fe）表面上的光生空穴聚集。空穴可以直接氧化有机分子，电子会使过氧化氢产生还原反应。通常，Bi_2WO_6/MIL-100（Fe）之间的 B 型异质结产生的内部电场促进了光生电子 - 空穴对的分离，从而增强了对水杨酸的光催化降解效率。

2.4 结论

总之，我们通过简便的溶剂热法成功地制备了一系列不同 MIL-100（Fe）含量的 Bi_2WO_6/MIL-100（Fe）纳米复合光催化材料。Bi_2WO_6/MIL-100（Fe）异质结结构在可见光下比纯 Bi_2WO_6 和 MIL-100（Fe）表现出更高的水杨酸光催化降解能力。当 MIL-100（Fe）质量分数为 3.5% 时，Bi_2WO_6/MIL-100（Fe）-2 的光催化活性达到最大。此外，Bi_2WO_6/MIL-100（Fe）纳米复合光催化材料在循环利用实验中显示出优异的光稳定性，这对实际应用非常重要。增强的光催化活性可以归因于 MIL-100（Fe）大的比表面积以及 Bi_2WO_6 和 MIL-100（Fe）之间形成的 p-n 结，其可以促进电子 - 空穴对的分离。

参考文献

[1] Zhao Y, Zhao B, Liu J, et al. oxide-modified nickel photocatalysts for the production of hydrocarbons in visible light[J]. Angew Chem Int Ed, 2016, 55(13): 4215-4219.

[2] Xiang Y, Ju P, Wang Y, et al. Chemical etching preparation of the Bi_2WO_6/BiOI p-n heterojunction with enhanced photocatalytic antifouling activity under visible light irradiation[J]. Chem Eng J,2016, 288: 264-275.

[3] Zhao Y, Jia X, Waterhouse G, et al. Ultrafine NiO nanosheets stabilized by TiO_2 from monolayer NiTi-LDH precursors: an active water oxidation electrocatalyst[J]. Adv Energy Mater, 2016, 6(6): 1501974-1501994.

[4] Yue D, Qian X, Zhao Y.Photocatalytic remediation of ionic pollutant[J]. Sci Bull, 2015, 60(21): 1791-1806.

[5] Shang L, Tong B, Yu H, et al.CdS nanoparticle-decorated Cd nanosheets for efficient visible light-driven photocatalytic hydrogen evolution[J]. Adv Energy Mater, 2016, 6(3): 1501241-1501247.

[6] Varley J, Janotti A, Van de Walle C.Mechanism of visible-light photocatalysis in nitrogen-

doped TiO$_2$[J]. Adv Mater, 2011, 23: 2343-2347.

[7]   Chandrasekaran S, Chung J, Kim E, et al.Exploring complex structural evolution of graphene oxide/ZnO triangles and its impact on photoelectrochemical water splitting[J]. Chem Eng J, 2016, 290: 465-476.

[8]   Tian G, Chen Y, Zhou W, et al.Facile solvothermal synthesis of hierarchical flower-like Bi$_2$MoO$_6$ hollow spheres as high performance visible-light driven photocatalysts[J]. J Mater Chem, 2011, 21: 887-892.

[9]   Monfort O, Pop L, Sfaelou S, et al.Comparative study between pristine and Nb-modified BiVO$_4$ films employed for photoelectrocatalytic production of H$_2$ by water splitting and for photocatalytic degradation of organic pollutants under simulated solar light[J]. Chem Eng J, 2016, 286: 91-97.

[10] Zheng Y, Duan F, Chen M, et al.Synthetic Bi$_2$O$_2$CO$_3$ nanostructures: Novel photocatalyst with controlled special surface exposed[J]. Mol Catal A: Chem, 2010, 317: 34-40.

[11] Liu Z, Ran H, Wu B, et al.Synthesis and characterization of BiOI/BiOBr heterostructure films with enhanced visible light photocatalytic activity[J]. Colloids Surfaces A: Physicochem Eng Aspects, 2014, 452: 109-114.

[12] Xu Y, Song J, Chen F, et al.Amorphous Ti(IV)-modified Bi$_2$WO$_6$ with enhanced photocatalytic performance[J]. RSC Adv, 2016, 6: 65902-65910.

[13] Duan F, Zhang Q, Shi D, et al.Enhanced visible light photocatalytic activity of Bi$_2$WO$_6$ via modification with polypyrrole[J]. Appl Surf Sci, 2013, 268: 129-135.

[14] Chen W, Kuang Q, Wang Q, et al.ChemInform abstract: Engineering a high energy surface of anatase TiO$_2$ crystals towards enhanced performance for energy conversion and environmental applications[J]. RSC Adv, 2015, 5: 20396-20409.

[15] Huang Z, Song J, Li K M, et al. Hollow cobalt-based bimetallic sulfide polyhedra for efficient all-pH-value electrochemical and photocatalytic hydrogen evolution[J]. J Am Chem Soc, 2016, 138: 1359-1365.

[16] Hu J, An W, Wang H, et al.Synthesis of a hierarchical BiOBr nanodots/Bi$_2$WO$_6$ p-n heterostructure with enhanced photoinduced electric and photocatalytic degradation performance[J]. RSC Adv, 2016, 6: 29554-29562.

[17] Chaiwichian S, Wetchakun K, Phanichphant S, et al.Highly efficient visible-light-induced photocatalytic activity of Bi$_2$WO$_6$/BiVO$_4$ heterojunction photocatalysts[J]. RSC Adv, 2016, 6: 54060-54068.

[18] Ju P, Wang Y, Sun Y, et al.Controllable one-pot synthesis of nest-like Bi$_2$WO$_6$/BiVO$_4$ composite with enhanced photocatalytic antifouling performance under visible light irradiation[J]. Dalton Trans, 2016, 45: 4588-4602.

[19] Jo W, Lee J, Natarajan T. Fabrication of hierarchically structured novel redox-mediator-free ZnIn$_2$S$_4$ marigold flower/Bi$_2$WO$_6$ flower-like direct Z-scheme nanocomposite photocatalysts with superior visible light photocatalytic efficiency[J]. Phys Chem Chem Phys, 2016, 18: 1000-1016.

[20] Yang W, Ma B, Wang W, et al.Enhanced photosensitized activity of a BiOCl-Bi$_2$WO$_6$ heterojunction by effective interfacial charge transfer[J]. Phys Chem Chem Phys, 2013, 15: 19387-19394.

[21] Bao J, Guo S, Gao J, et al.Synthesis of Ag$_2$CO$_3$/Bi$_2$WO$_6$ heterojunctions with enhanced photocatalytic activity and cycling stability[J]. RSC Adv, 2015, 5: 97195-97204.

[22] Jiang H, Makal T, Zhou H.Interpenetration control in metal-organic frameworks for functional applications[J]. Coord Chem Rev, 2013, 257: 2232-2249.

[23] Murray L, Dincǎ M, Long J.Hydrogen storage in metal-organic frameworks[J]. Chem Soc Rev, 2009, 38: 1294-1314.

[24] Lee J, Farha O, Roberts J, et al.Metal-organic framework materials as catalysts[J]. Chem Soc Rev, 2009, 38: 1450-1459.

[25] Zhan W, Kuang Q, Zhou J, et al.Semiconductor@ metal-organic framework core-shell heterostructures: A case of ZnO@ ZIF-8 nanorods with selective photoelectrochemical response[J]. J Am Chem Soc, 2013, 135: 1926-1933.

[26] Zhan W, He Y, Guo J, et al.Probing the structural flexibility of MOFs by constructing metal oxide@ MOF-based heterostructures for size-selective photoelectrochemical response[J]. Nanoscale, 2016, 8: 13181-13185.

[27] Zhang M, Li B, Li Y, et al.Finely tuning MOFs towards high performance in C$_2$H$_2$ storage: Synthesis and properties of a new MOF-505 analogue with an inserted amide functional group[J]. Chem Commun, 2016, 52: 7241-7244.

[28] Alvaro M, Carbonell E, Ferrer B, et al. Semiconductor behavior of a metal-organic framework (MOF)[J].Chem A Eur J, 2007, 13: 5106-5112.

[29] Tachikawa T, Choi J, Fujitsuka M, et al.Photoinduced charge-transfer processes on MOF-5 nanoparticles: Elucidating differences between metal-organic frameworks and semiconductor metal oxides[J]. J Phys Chem C, 2008, 112: 14090-14101.

[30] Wang J, Wang C, Lin W.Metal-organic frameworks for light harvesting and photocatalysis[J]. ACS Catal, 2012, 2: 2630-2640.

[31] Nasalevich M, Goesten M, Savenije T, et al.Enhancing optical absorption of metal-organic frameworks for improved visible light photocatalysis[J]. Chem Commun, 2015, 51: 961-962.

[32] Nasalevich M, Becker R, Ramos-Fernandez E, et al.Co@ NH$_2$-MIL-125(Ti): Cobaloxime-derived metal-organic framework-based composite for light-driven H$_2$ production[J]. Energy Environ Sci, 2015, 8: 364-375.

[33] Laurier K, Vermoortele F, Ameloot R, et al.Iron(III)-based metal-organic frameworks as visible light photocatalysts[J]. J Am Chem Soc, 2013, 135(2013): 14488-14491.

[34] Ke F, Wang L, Zhu J.Facile fabrication of CdS-metal-organic framework nanocomposites with enhanced visible-light photocatalytic activity for organic transformation[J]. Nano Res, 2015, 8: 1834-1846.

[35] Zhou J, Wang R, Liu X, et al.In situ growth of CdS nanoparticles on UiO-66 metal-organic framework octahedrons for enhanced photocatalytic hydrogen production under visible light irradiation[J]. Appl Surf Sci, 2015, 346: 278-283.

[36] Rad M, Dehghanpour S.Discrete molecular complex, one and two dimensional coordination polymer from cobalt, copper, zinc and (E)-4-hydroxy-3-((quinolin-8-ylimino)methyl)benzoic acid: Synthesis, structures and gas sensing property[J]. RSC Adv, 2016, 6: 61784-61793.

[37] Zhang J, Zhang H, Du Z, et al.Water-stable metal-organic frameworks with intrinsic peroxidase-like catalytic activity as a colorimetric biosensing platform[J]. Chem Commun, 2014, 50: 1092-1094.

[38] Ma J, Song W, Chen C, et al.Fenton degradation of organic compounds promoted by dyes under visible irradiation[J]. Environ Sci Technol, 2005, 39: 5810-5815.

[39] Zhang C, Qiu L, Ke F, et al.A novel magnetic recyclable photocatalyst based on a core-shell metal-organic framework Fe_3O_4@MIL-100(Fe) for the decolorization of methylene blue dye[J]. J Mater Chem A, 2013, 1: 14329-14334.

[40] Xu Q, Wellia D, Ng Y, et al.Synthesis of porous and visible-light absorbing Bi_2WO_6/TiO_2 heterojunction films with improved photoelectrochemical and photocatalytic performances[J]. J Phys Chem C, 2011, 115: 7419-7428.

[41] Horcajada P, Surblé S, Serre C, et al.Synthesis and catalytic properties of MIL-100(Fe), an iron(Ⅲ) carboxylate with large pores[J]. Chem Commun, 2007, 27: 2820-2822.

[42] Ma Y, Chen Z, Qu D, et al.Synthesis of chemically bonded BiOCl@Bi_2WO_6 microspheres with exposed (020) Bi_2WO_6 facets and their enhanced photocatalytic activities under visible light irradiation[J]. Appl Surf Sci, 2016, 361: 63-71.

[43] Zhang C, Ai L, Jiang J.Graphene hybridized photoactive iron terephthalate with enhanced photocatalytic activity for the degradation of Rhodamine B under visible light[J].Ind Eng Chem Res, 2014, 54: 153-163.

[44] He Z, Shi Y, Gao C, et al.BiOCl/$BiVO_4$ p-n heterojunction with enhanced photocatalytic activity under visible-light irradiation[J]. J Phys Chem C, 2013, 118: 389-398.

[45] Shamaila S, Sajjad A, Chen F, et al.Mesoporous titania with high crystallinity during synthesis by dual template system as an efficient photocatalyst[J]. J Colloid Interface Sci, 2011, 356: 465-472.

第三章

BiOI/磁性树脂复合光催化材料制备及性能研究

3.1 引言	97
3.2 实验部分	101
3.3 结果与讨论	103
3.4 结论	108
参考文献	108

3.1 引言

废水主要含有药品、染料、农药等工业化学品生产过程中产生的有机污染物，会对自然环境资源产生严重危害，并对人类健康造成不利影响。其中，1-氨基-8-萘酚-3,6-二磺酸（H 酸）是合成染料的重要中间体之一，它主要用于生产染料，也可用于制药工业。H 酸是以萘为原料，在一定条件下，经磺化、硝化、中和、还原、碱溶和酸析等化学过程制取的。H 酸生产工艺流程长，原料利用率低，生产过程中排出的废液往往含有大量萘的各种取代衍生物，颜色深、酸性强。由于废液中的有机物大多是带有硝基、氨基和磺酸基等基团的芳香族化合物，所以对微生物有强烈的毒性。带有磺酸基的芳香族化合物易溶于水，传统的物理化学方法处理效率很低。到现在为止，由于 H 酸分子存在稠环系统，很难通过物理化学方法处理 H 酸废水[1-2]，所以，H 酸废水已成为环境污染控制的难点和热点。

20 世纪 90 年代以来，随着学科之间不断交叉，光催化技术在环境保护、能源开发、有机合成等方面的应用研究发展迅速，成为国际上最活跃的研究领域之一。当用能量等于或大于半导体禁带宽度（带隙能，E_g）的光照射半导体时，半导体价带的电子跃迁到导带，在导带上产生带负电的高活性电子（e^-），在价带上留下带正电荷的空穴（h^+），这样就形成光生电子-空穴，到达半导体粒子表面的电子和空穴能够分别进行两个反应。电子能够还原被吸附的电子受体，在富氧的溶液中，通常电子被溶解氧吸收还原为氧自由基，空穴可以氧化表面吸附的电子供体。根据热力学原理，光催化氧化还原反应要求受体电势比半导体导带电势低（更正）且供体电势比半导体价带电势高（更负），才能进行有效反应[3-5]。近年来，具有可见光活性的铋系列复合氧化物已发展成为光催化领域的一个研究焦点。铋系列的光催化剂一般为 $A_xB_yO_z$ 型结构，其价带由 Bi6s 和 O2p 轨道杂化组成，Bi6s 轨道与 O2p 轨道的强相互作用降低了其对称性，使其具有较高的电荷流动性和氧化活性，并且可见光吸收是其本身的带间跃迁，不存在复合中心的说法，从而具有较高的光催化活性，另外，铋系列光催化剂一般都具有独特的层状结构，在层间进行的"二维"光催化反应会随层间分子或离子的改变而改变。

铋系列光催化剂主要包括：Bi_2WO_6、$Bi_2Ti_2O_7$、$BiPO_4$、BiOX（X=F，Cl，Br，I）、$BiVO_4$、Bi_2MoO_6、Bi_2FeO_4 和 Bi_3NbO_7 等。其中，Bi_2WO_6、$BiVO_4$、BiOX（X=F，Cl，Br，I）和 Bi_2MoO_6 这4类含铋复合氧化物由于具有较好的光催化活性，是目前铋系列光催化剂中研究最广泛的。碘氧化铋英文名称：bismuth oxyiodide；分子式：BiOI；性质：红色结晶四方晶体；密度：$7.922g/cm^3$。碘氧化铋溶于盐酸，不溶于水、醇、氯仿，遇硝酸或碱则分解，加热至红热时熔融而部分分解，溶解于盐酸变为黄色；在空气中加热变成氧化铋，在隔绝空气下加热时部分分解升华；由碘化铋与沸腾水进行长时间作用制得，亦可用氧化铋溶于氢碘酸，加入过量的水加热反应制得，还可用碱式碳酸铋与热氢碘酸反应制得。

在许多的可见光光催化剂当中，因卤氧化铋BiOX（X=F，Cl，Br，I）结构独特和光催化性能良好[6-10]，卤氧化铋BiOX（X=F，Cl，Br，I）占有很重要的地位，其Bi 6s轨道可以通过与O2p轨道杂化使带隙变窄，而$[Bi_2O_2]^{2+}$层和X^-层交替的结构使其非常容易长成片状结构，且$[Bi_2O_2]^{2+}$层和X^-层之间可形成内建电场，可促进光生电子和空穴分离，从而其光催化活性得以增强。在卤氧化铋BiOX（X=F，Cl，Br，I）中，BiOI的带隙最窄（E_g=1.8～1.9eV），其吸收带边波长可到680nm，吸收光的范围几乎包括全部可见光（390～760nm），可见光响应性能良好。但是，因为BiOI块体的光生电子和空穴容易复合[11-12]，它的光催化活性无法得到充分利用。众多研究者试图通过形貌与结构调控、异质结构建及表面修饰等方法对其性质进行改善[13-15]，期望其在可见光范围内光催化性能得以最大限度发挥。其中用制备复合光催化剂来处理这个问题，是一个行之有效的策略。到目前为止，许多报道已经声明，许多以BiOI为基础制备的复合材料，其光催化活性得到极大增强[16-25]，如BiOI/BiOBr、BiOI@$Bi_{12}O_{17}Cl_2$、$BiOI/BiVO_4$、$Bi_2WO_6/BiOI$、$BiOI/TiO_2$、$BiOI/Fe_3O_4$、BiOI/ZnO、$BiOI/BiPO_4$、$BiOI/Bi_4Ti_3O_{12}$ 和 $g-C_3N_4/Fe_3O_4/BiOI$ 等材料的光催化活性就得到明显的增强。这些结果表明，BiOI和其他半导体对应的能带隙相耦合，这实际上能促进电荷分离，从而提高其光催化性能。

离子交换树脂的定义：离子交换树脂是一类带有可离子化基团的三

维网状高分子材料,其外形一般为颗粒状。不溶于水和一般的酸、碱,也不溶于普通的有机溶剂(如烃类溶剂)。常见的离子交换树脂的粒径为 0.3~1.2nm。

离子交换树脂的结构:①三维空间结构的网架;②骨架上连接的可离子化的功能基团;③功能基团上吸附的可交换的离子材料。

离子交换的原理:离子交换过程是在水溶液与固体交换剂之间的被分离组分(即被提取、被纯化的离子或分子)所发生的一种化学计量分配过程,该过程遵循固 - 液非均相扩散传质的普遍规律,但不同于传统分离过程。当被分离组分与溶液接触时,离子交换剂会与溶液中的特定离子进行交换,即离子交换树脂上的可交换离子(阳离子或阴离子)被溶液中带同种电荷的特定离子所取代,而不溶性固体骨架在这一交换过程中不会发生任何化学变化。该过程一般可以用方程式表达为:R—B+A$^+$ ⟶ R—A+B$^+$ (R代表树脂中除可交换离子以外的其他部分,即惰性骨架与固定基团;B为可交换离子;A为待分离组分)。

离子交换树脂的分类如下。

按交换基团的性质分类:阳离子交换树脂分为强酸型、中酸型、弱酸型;阴离子交换树脂分为强碱型、弱碱型。阴离子交换树脂含有季铵基[—N(CH$_3$)$_3$OH]、氨基(—NH$_2$)或亚氨基(—NH—)等碱性基团。在水中它们能生成 OH$^-$,可与各种阴离子交换,其交换原理为:

$$R—N(CH_3)_3OH + Cl—R \longrightarrow N(CH_3)_3Cl + OH^-$$

按树脂的物理结构分类:凝胶型、大孔型、载体型。大孔型离子交换树脂外观不透明,表面粗糙,为非均相凝胶结构。即便在干燥状态,大孔型离子交换树脂内部也存在尺寸不同的毛细孔,因此可在非水体系中起吸附和离子交换作用,其比表面积大,因此吸附功能十分显著。离子交换树脂在废水、废气的浓缩、处理、分离、回收及分析检测上都有重要应用,普遍运用于电镀废水、造纸废水、矿冶废水、生活污水、影片洗印废水、工业废气等的治理。例如影片洗印废水中的银是以 Ag(SO$_3$)$_2^{3-}$ 等阴离子形式存在的,经过Ⅰ型强碱型离子交换树脂处理后,银的回收率可达 90% 以上,既节约了成本,又使废水的排放达标。又如电镀废水中含有大量有毒的金属氰化物，如 $Fe(CN)_6^{3-}$、$Fe(CN)_6^{4-}$ 等，用抗有机污染力强的阴离子交换树脂处理后，可使金属氰化物的含量大幅下降，如经过聚丙烯酰胺系阴离子交换树脂处理后，金属氰化物的含量降至 10mg/L 以下。磁性树脂是澳大利亚联邦科学与工业研究院、澳瑞凯（Orica）公司和南澳水务局共同研发的，用来去除水中有机物的新型阴离子交换树脂。这种树脂是以聚丙烯为母体的季铵型离子交换树脂，通过可交换氯离子（Cl^-）与水中带负电的天然有机物（NOM）进行离子交换，从而达到净水目的。与传统离子交换树脂相比，这类树脂对有机物交换容量大、去除效率高，且由于树脂颗粒内部含磁核，可使树脂相互聚集成团实现快速沉淀。因此，该树脂与其他离子交换树脂不同，在搅拌式反应器中可进行吸附，从而改变了吸附柱式的吸附模式，应用前景广阔[26-27]。磁性聚丙烯酸阴离子交换树脂（PAER）是一种有着大孔结构的新型离子交换树脂，它因为体积小而具有快速吸附能力，并能根据有机污染物不同的化学性质而去除污染物，这表明它在废水处理中有着巨大潜力[21-22]。PAER 树脂存在磁化成分（Fe_3O_4）[28]，这使其在磁力搅拌过程中能够自发地聚集和分离[23]。因此，树脂吸附法以其操作简单、效率高、成本低廉等优点，而被认为是最有效的处理废水的技术之一。然而，由于 PAER 制备困难且再生吸附材料价格昂贵，它能否循环利用就成为了实际应用中至关重要的问题。

依据现有的报道，很少有 BiOI 和 PAER 树脂结合重组的实验案例。基于以上思路，本实验通过低温共沉淀法合成一组具有可见光响应的复合光催化剂——碘氧化铋/磁性聚丙烯酸阴离子交换树脂（BiOI/PAER），来降解有机污染物。经过光催化剂在可见光照下降解染料的中间体 1-氨基-8-萘酚-3,6-二磺酸（H 酸），筛选出性能最优的 BiOI/PAER 复合光催化剂质量配比，并通过各种仪器表征分析，评估其光催化性能。结果发现 1h 内在 BiOI/PAER-2 光催化剂作用下 1-氨基-8-萘酚-3,6-二磺酸的降解率达到 90.1%，明显高于 1h 内在纯 BiOI 光催化剂作用下 1-氨基-8-萘酚-3,6-二磺酸的降解率（50.3%）。这是由于在 BiOI 和磁性树脂中的 Fe_3O_4 之间形成异质结有助于光生电荷载体的转移和分离，BiOI/PAER 复合材料的光催化性能得到增强，除此之外，BiOI/PAER 复合材料具有优良的稳定性和可分离性（图 3-1）。

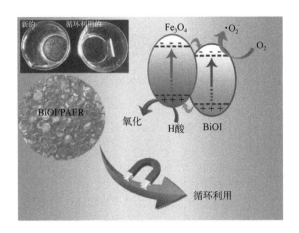

图 3-1　在可见光照射下 BiOI/PAER 复合材料的催化作用的机理

3.2　实验部分

3.2.1　BiOI/聚丙烯酸阴离子交换树脂复合材料的合成

通过低温共沉淀制备 BiOI/聚丙烯酸阴离子交换树脂复合材料。首先，往去离子水中加入硝酸溶液（HNO_3）并不断搅拌，调得溶液 pH=2.0，称取 $Bi(NO_3)_3 \cdot 5H_2O$（0.388g，0.8mmol），将其溶解在 100mL 的该溶液中得到 $Bi(NO_3)_3$ 溶液。其次，称取 20、50、100、150mg 的 PAER 和 KI（0.0528g，0.3mmol）分别加到 30mL 的去离子水中，剧烈搅拌获得均匀分散的 PAER-KI 混合液。然后，将 $Bi(NO_3)_3$ 溶液逐滴加入 PAER-KI 混合液中。并往该混合液中添加氨水，调整溶液到 pH = 3，随后加热至 85℃，连续搅拌 24h。通过磁力搅拌分离得到的产品，经乙醇和水多次洗涤，在 40℃真空下干燥 12h。最终得到 BiOI/PAER 复合光催化产品，其中 BiOI 和 PAER 的质量比分别为 1∶40、1∶20、1∶10 和 1∶5，可以记为 BiOI/PAER-1、BiOI/PAER-2、BiOI/PAER-3 和 BiOI/PAER-4。用同样的方法合成没有引入磁性聚丙烯酸阴离子交换树脂（PAER）的纯 BiOI，作为空白对照。该实验的主要步骤及技术路线如图 3-2 所示。

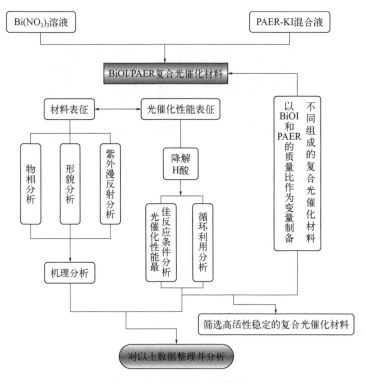

图 3-2 实验的技术路线图

3.2.2 实验取样分析

在室温下，通过可见光（300W 氙灯，$\lambda > 420$nm）照射降解 H 酸，评估 BiOI/PAER 复合材料的光催化性能。在实验装置中，将 50mg 催化剂加入 100mL 的 H 酸溶液（c=200mg/L）中。照射前，悬浮液要在暗室中用电动搅拌器连续搅拌 15h 来达到复合光催化剂和 H 酸之间的吸附 - 解吸平衡。每隔 30min，从悬浮装置中迅速收集约 3mL 的等分试样并离心，取 2mL 上清液。

用高效液相色谱法（Waters 2695）测定试样的 H 酸浓度。用 C18 分析柱（150mm×4.6mm，填料粒径 5μm）对样品进行分离。把含有 0.5%KH_2PO_4 甲醇 - 水（体积分数为 50%）当作 pH 调节剂的流动相，流动相的流速为 0.8cm^3/min。在光照前和可见光照射不同时间之后，使用岛津 UV2501PC 光谱仪分别测定 H 酸溶液的紫外 - 可见光谱。并用重铬酸钾法测定化学需氧量（COD），用燃烧氧化 - 非色散红外吸收法（上海元析仪器有限公司 TOC-2000）测定 H 酸的总有机碳量（TOC）。

3.2.3 表征

由布鲁克 AXS D8 Advancex X 射线衍射仪和 CuKα 射线（λ=0.15406nm）测定所制备的产品的 X 射线衍射图谱。用 JEOL Hitachi S-4800 型场发射扫描电子显微镜（FESEM）进行扫描得到 SEM 图像。由 JEOL JEM-2100 显微镜获得高分辨透射电子显微镜图像。用紫外 - 可见光谱仪测定样品的紫外 - 可见漫反射光谱并与 $BaSO_4$ 的紫外 - 可见漫反射光谱相比较。用比表面积分析仪计算样品的比表面积（美国麦克仪器，ASAP2020）。

3.3 结果与讨论

图 3-3 是 BiOI、PAER 和 BiOI/PAER-2 复合材料的 X 射线衍射光谱，从图中可以看出，BiOI 所有衍射峰吻合于四方相（JCPDS 73-2062）。图中无杂质峰，说明 BiOI 产品纯度高。此外，磁性聚丙烯酸阴离子交换树脂的特征峰明显地对应于 Fe_3O_4（JCPDS 00-075-0033）的立方结构。在 BiOI/PAER-2 复合材料的 XRD 图谱中，特征峰的出现可以归因于四方相的 BiOI 和立方结构的 Fe_3O_4。在 BiOI/PAER-2 复合材料的 X 射线衍射光谱中无杂质峰，表明复合材料仅由 BiOI 和磁性聚丙烯酸阴离子交换树脂所制备。

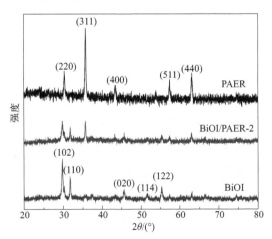

图 3-3　BiOI、PAER 和 BiOI/PAER-2 复合材料的 X 射线衍射光谱

通过对所制备材料的 SEM、TEM 和 HRTEM 图像进行分析研究，我们得出了样品的显微结构和形态。如图 3-4（a）和图 3-4（c）所示，树脂颗粒表面

粗糙，它的颗粒直径分布在 15μm 左右。而且，树脂内有很多大小约为 0.5μm 的孔道。很明显图 3-4（b）是 BiOI 的 TEM 图像，纯 BiOI 由尺寸小于 0.1μm 的不规则平滑纳米片组成，而从图 3-4（c）中观察，可发现 BiOI/PAER-2 复合材料中不规则滑片仍然存在，并且它附着在树脂的表面。图 3-4（d）显示小滑片 BiOI 每隔 0.298 nm 条纹统一，这与四方晶格面（102）相一致。

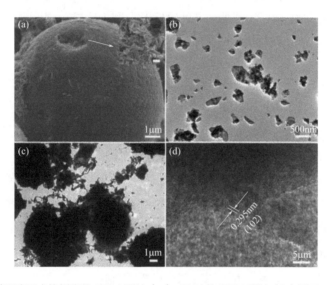

图 3-4　磁性阴离子交换树脂的 SEM 图片（a）、BiOI 的 TEM 图片（b）以及 BiOI/PAER-2 复合材料的 TEM 和 HRTEM 图片（c, d）

此外，在 77K 的低温氮吸附-脱附等温线下，对 BiOI、PAER 和 BiOI/PAER-2 的样品进行比表面积测试（BET），其比表面积分别是 16.1619m^2/g、2.3216m^2/g 和 2.3390m^2/g。其中 BiOI/PAER-2 的 BET 比表面积与纯 PAER 的 BET 比表面积相似。因此，可推测 BiOI/PAER-2 也具有快速吸附能力。

带隙能是决定光催化活性的重要因素之一。仔细观察所制备样品的紫外-可见漫反射光谱（DRS）[图 3-5（a）]，可以看出在 BiOI 和 PAER 端光谱分别约在 650nm、720nm 处的光线基本被吸收，这表明 BiOI/PAER 复合材料在可见光区有较强的吸收能力，与 PAER 类似，它的最大吸收波长也在 720nm 左右。结果如图 3-5（b）所示，光谱曲线的切线与横坐标的交点为 $(\alpha h\upsilon)^{1/2}$ 与光子能量的带隙能量。纯 BiOI 和 PAER 带隙能量分别是 1.95eV 和 1.69eV。而 BiOI/PAER-1~BiOI/PAER-4 的带隙能量分别是 1.74eV、1.73eV、1.75eV 和

1.72eV。BiOI/PAER 复合材料的带隙能量在纯 BiOI 和 PAER 的带隙能量之间。这些结果表明，在可见光辐照下所制备的 BiOI/PAER 复合材料的光催化剂是很有前景的。

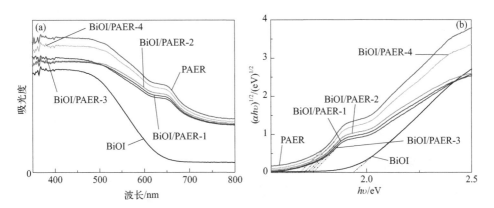

图 3-5　所制备的样品的紫外－可见漫反射吸收光谱（a）和测试 $(\alpha h\upsilon)^{1/2}$ 与光子能之间的带隙能量（b）

如图 3-6（a）所示，随着 H 酸浓度增加，BiOI/PAER-2 复合材料的吸附量几乎不变，它的光催化活性同样变化不大。如图 3-6（b）所示，由于 H 酸分子的电离加剧，BiOI/PAER-2 的吸附量随 pH 值的增加而增加，但 BiOI/PAER-2 光催化活性随 pH 变化不大。如图 3-6（c）所示，在可见光照射相同时间下，不同光催化剂对 H 酸的光降解率不同。从图 3-6（c）可以看出，无光的 H 酸是相当稳定的。在没有光催化剂的情况下，可见光照射 210min 内 H 酸的自降解率为 20%；但相同条件下加入 BiOI，210min 内 H 酸的自降解率为 80%。就纯 PAER 和 BiOI/PAER 复合材料而言，开始无光的 15h 内，H 酸的浓度大约减少了 50%，这是由于 H 酸吸附 PAER。之后，该废水浓度保持不变，这表明树脂达到了吸附 H 酸的平衡。在可见光照射下，由于 H 酸的自降解，其浓度逐渐下降，50min 内约下降 11%。而使用 BiOI/PAER 复合材料 150min 内的 H 酸降解率约 90%，BiOI/PAER 复合材料的光催化活性与 BiOI、PAER 相比明显增强。这些 BiOI/PAER 复合材料中，BiOI/PAER-2 对 H 酸降解的光催化活性最高。在相同的情况下，H 酸 60min 内降解 90%。然而，在那之后，在相同的条件下，H 酸废水浓度保持不变，所以它不可能只通过光降解达到 100% 的。溶液降解前后的化学需氧量（COD）经测定分别是 186.48

mg/L 和 12.60mg/L，这与高效液相色谱法测定 H 酸浓度提供的数据相一致。同样溶液降解前后的总有机碳量（TOC）分别是 61.4824mg/L 和 6.3503mg/L。图 3-6（d）是光催化剂 BiOI/PAER-2 降解不同时间后 H 酸溶液的紫外 - 可见吸收光谱。光降解的过程中，随着反应时间的增加，H 酸的吸收峰强度逐渐降低。在可见光照射 60min 内，BiOI/PAER-2 复合材料可降解大部分的 H 酸（90%）。结果表明，PAER 和 BiOI 组合能显著提高光催化活性。

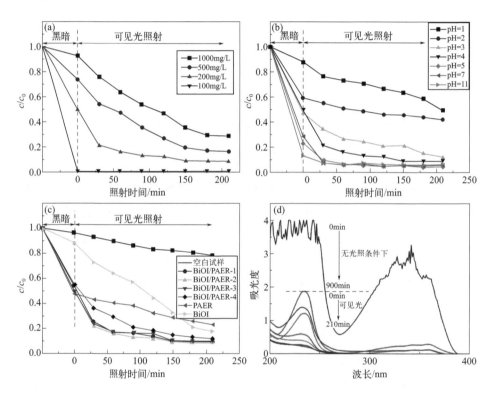

图 3-6　H 酸浓度对 BiOI/PAER-2 复合材料降解 H 酸废水的影响（pH=5）（a）；H 酸 pH 值对 BiOI/PAER-2 复合材料降解 H 酸的影响（$c_{H酸}$=200mg/L）（b）；在可见光下照射下，不同光催化剂降解 H 酸的性能比较（pH=5，$c_{H酸}$=200mg/L）（c）；在暗室和可见光下照射不同时间长度的 H 酸溶液，随 BiOI/PAER-2 复合材料的催化降解的紫外 - 可见光谱（d）

图 3-7 展示的是 BiOI/PAER-2 复合材料五次循环利用前后可见光照射下 H 酸的 XRD 图谱。很明显，使用过的 BiOI/PAER-2 复合材料与未使用过的 BiOI/PAER-2 复合材料的 XRD 图谱类似，可以证明在 H 酸溶液中的光降解

催化剂 BiOI/PAER-2 的化学性质稳定。BiOI/PAER 复合材料能简单磁性再生，这在实际应用中是很重要的。这些结果表明，BiOI/PAER 复合材料在可见光光催化剂的实际应用领域会大有作为。从实际应用来看，光催化剂的稳定循环和再生循环利用具有重要意义。用可见光光催化剂降解 H 酸，五次磁力循环使用后，BiOI/PAER-2 复合材料的在 50min 内光催化效率从 90% 下降到 74%。结果表明，所制备的 BiOI/PAER 复合材料五次循环利用后在可见光波段仍稳定。

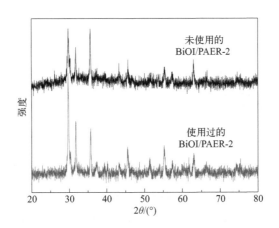

图 3-7　未使用和使用过的 BiOI/PAER-2 复合材料样品 XRD 图谱

BiOI/PAER 的光催化性能优良，也许能被归结为以下几点原因。首先，PAER 提高了吸附 H 酸的能力，并使 H 酸分子富集在 BiOI/PAER 复合材料表面，从而能够加速光催化反应。其次，在 PAER 中 BiOI 和 Fe_3O_4 之间的异质结构促进了光生电子-空穴对的分离和局部空间氧化，并有效提高了 H 酸光催化降解效率。如图 3-8 所示，展示的是在 BiOI/PAER 复合材料作用下光催化降解 H 酸的机理。Fe_3O_4 和 BiOI 的带隙能量分别是 0.10eV 和 1.72eV。Fe_3O_4 和 BiOI 的导带（CB）末端分别是 0.17eV 和 0.58eV。相应地，Fe_3O_4 和 BiOI 的价带（VB）顶端分别是 0.27eV 和 2.30eV[29-31]。当复合材料在可见光照射下时，光生电子-空穴对会出现在 Fe_3O_4 和 BiOI 表面或体内。同时，电子可以从 Fe_3O_4 的导带（CB）迁移到 BiOI 的导带（CB），而空穴可以从 BiOI 的价带（VB）转移到 Fe_3O_4 的价带（VB）。其结果是，可以有效地抑制电子-空穴对的重组，这解释了复合材料促进降解过程的原因。

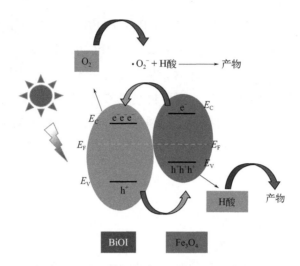

图 3-8　在可见光照射下，BiOI/PAER 复合材料催化作用的机理及电荷分离过程图示

3.4　结论

总之，通过 BiOI 纳米粒子和磁性聚丙烯酸阴离子交换树脂的复合，成功制备了复合 BiOI/PAER 异质结构，该催化剂对现实生活中的主要污染物之一 H 酸有良好的光降解效应。BiOI/PAER 复合材料具有吸附容量大，在可见光下有良好的光催化活性的优点，这可能是因为电荷分离效率的增强。BiOI/PAER 复合材料在循环实验中有着优良的光稳定性，这会在实际应用中成为一个关键因素。这些结果表明，在这项实验中合成的 BiOI/PAER 复合材料，在工业废水处理方面具有广阔的应用前景。

参考文献

[1]　Mohanty S, Rao N, Khare P, et al.A coupled photocatalytic-biological process for degradation of 1-amino-8-naphthol-3,6-disulfonic acid (H-acid)[J].Water Res, 2005, 39: 5064-5070.

[2]　Zhu W, Yang Z, Wang L.Application of ferrous-hydrogen peroxide for the treatment of H-acid manufacturing process wastewater[J].Water Res, 1996, 30: 2949-2954.

[3]　Shang L, Tong B, Yu H, et al.CdS nanoparticle-decorated Cd nanosheets for efficient visible light-driven photocatalytic hydrogen evolution[J].Adv Energy Mater, 2016, 6: 1501241-

1501247.

[4] Mehraj O, Pirzada B, Mir N, et al.A highly efficient visible-light-driven novel p-n junction Fe_2O_3/BiOI photocatalyst: Surface decoration of BiOI nanosheets with Fe_2O_3 nanoparticles[J]. Appl Surf Sci, 2016, 387: 642-651.

[5] Jiao Z, Zheng J, Feng C, et al.Co-doped $BiVO_4$ photoanodes with a metal-organic framework cocatalyst for improved photoelectrochemical stability and activity[J].ChemSusChem, 2016, 9: 2824-2831.

[6] He T, Wu D, Tan Y.Fabrication of BiOI/$BiVO_4$ heterojunction with efficient visible-light-induced photocatalytic activity[J].Mater Lett, 2016, 165: 227-230.

[7] Song H, Wu R, Liang Y, et al.Facile synthesis of 3D nanoplate-built $CdWO_4$/BiOI heterostructures with highly enhanced photocatalytic performance under visible-light irradiation[J].Colloids Surf A, 2017, 522: 346-354.

[8] Malathi A, Arunachalam P, Grace A, et al.A robust visible-light driven $BiFeWO_6$/BiOI nanohybrid with efficient photocatalytic and photoelectrochemical performance[J].Appl Surf Sci, 2017, 412: 85-95.

[9] Zhou R, Wu J, Zhang J, et al.Photocatalytic oxidation of gas-phase HgO on the exposed reactive facets of BiOI/$BiOIO_3$ heterostructures[J].Appl Catal B, 2017, 204: 465-474.

[10] Yosefi L, Haghighi M, Allahyari S.Solvothermal synthesis of flowerlike p-BiOI/n-$ZnFe_2O_4$ with enhanced visible light driven nanophotocatalyst used in removal of acid orange 7 from wastewater[J].Sep.Purif Technol, 2017, 178: 18-28.

[11] Lei Y, Wang G, Song S, et al.Room temperature, template-free synthesis of BiOI hierarchical structures: Visible-light photocatalytic and electrochemical hydrogen storage properties[J]. Dalton Trans, 2010, 39: 3273-3278.

[12] Xia J, Yin S, Li H, et al.Self-Assembly and enhanced photocatalytic properties of BiOI hollow microspheres via a reactable ionic liquid[J].Langmuir, 2010, 27: 1200-1206.

[13] Bao J, Guo S, Gao J, et al.Synthesis of Ag_2CO_3/Bi_2WO_6 heterojunctions with enhanced photocatalytic activity and cycling stability[J].RSC Adv, 2015, 5: 97195-97204.

[14] Tian J, Sang Y, Yu G, et al.A Bi_2WO_6-based hybrid photocatalyst with broad spectrum photocatalytic properties under UV, visible, and near-infrared irradiation[J].Adv Mater, 2013, 25: 5075-5080.

[15] Zhou J, Wang R, Liu X, et al.In situ growth of CdS nanoparticles on UiO-66 metal-organic framework octahedrons for enhanced photocatalytic hydrogen production under visible light irradiation[J].Appl Surf Sci, 2015, 346: 278-283.

[16] Liu Z, Ran H, Wu B, et al.Synthesis and characterization of BiOI/BiOBr heterostructure films with enhanced visible light photocatalytic activity[J].Colloids Surf A, 2014, 452: 109-114.

[17] Huang H, Xiao K, He Y, et al.In situ assembly of BiOI@$Bi_{12}O_{17}Cl_2$ p-n junction: Charge induced unique front-lateral surfaces coupling heterostructure with high exposure of BiOI {001} active facets for robust and nonselective photocatalysis[J].Appl Catal B, 2016, 199: 75-86.

[18] Ye K, Chai Z, Gu J, et al.BiOI-$BiVO_4$ photoanodes with significantly improved solar water splitting capability: P-n junction to expand solar adsorption range and facilitate charge carrier dynamics[J].Nano Energy, 2015, 18: 222-231.

[19] Xiang Y, Ju P, Wang Y, et al.Chemical etching preparation of the Bi_2WO_6/BiOI p-n heterojunction with enhanced photocatalytic antifouling activity under visible light irradiation[J].Chem Eng J, 2016, 288: 264-275.

[20] Dai G, Yu J, Liu G.Synthesis and enhanced visible-light photoelectrocatalytic activity of p-n junction BiOI/TiO_2 nanotube arrays[J].J Phys Chem C, 2011, 115: 7339-7346.

[21] Li X, Niu C, Huang D, et al.Preparation of magnetically separable Fe_3O_4/BiOI nanocomposites and its visible photocatalytic activity[J].Appl Surf Sci, 2013, 286: 40-46.

[22] Jiang J, Zhang X, Sun P, et al.ZnO/BiOI heterostructures: Photoinducedcharge-transfer property and enhanced visible-light photocatalytic activity[J].J Phys Chem C, 2011, 115: 20555-20564.

[23] Cao J, Xu B, Lin H, et al.Highly improved visible light photocatalytic activity of $BiPO_4$ through fabricating a novel p-n heterojunction BiOI/$BiPO_4$ nanocomposite[J].Chem Eng J, 2013, 228: 482-488.

[24] Hou D, Hu X, Hu P, et al.$Bi_4Ti_3O_{12}$ nanofibers-BiOI nanosheets p-n junction: Facile synthesis and enhanced visible-light photocatalytic activity[J].Nanoscale, 2013, 5: 9764-9772.

[25] Mousavi M, Habibi-Yangjeh A.Magnetically separable ternary g-C_3N_4/Fe_3O_4/BiOI nanocomposites: Novel visible-light-driven photocatalysts based on graphitic carbon nitride[J].J Colloid Interface Sci, 2016, 465: 83-92.

[26] Jarvis P, Mergen M, Banks J, et al.Pilot scale comparison of enhanced coagulation with

magnetic resin plus coagulation systems[J].Environ Sci Technol, 2008, 42: 1276-1282.

[27] Neale P, Schäfer A.Magnetic ion exchange: Is there potential for international development?[J].Desalination, 2009, 248: 160-168.

[28] Shuang C, Pan F, Zhou Q, et al.Magnetic polyacrylicanion exchange resin: Preparation, characterization and adsorption behavior of humic acid[J].Ind Eng Chem Res, 2012, 51: 4380-4387.

[29] Liu S.Magnetic semiconductor nano-photocatalysts for the degradation of organic pollutants[J].Environ Chem Lett, 2012, 10: 209-216.

[30] Cao J, Xu B, Luo B, et al.Novel BiOI/BiOBr heterojunction photocatalysts with enhanced visible light photocatalytic properties[J].Catal Commun, 2011, 13: 63-68.

[31] Xiao X, Hao R, Liang M, et al.One-pot solvothermal synthesis of three-dimensional (3D) BiOI/BiOCl composites with enhanced visible-light photocatalytic activities for the degradation of bisphenol-A[J].J Hazard Mate, 2012, 233: 122-130.

第四章

UiO-66/BiOI 复合光催化材料制备及性能研究

4.1	引言	113
4.2	实验部分	117
4.3	结果与讨论	119
4.4	结论	124
	参考文献	125

4.1 引言

目前,能源短缺、环境污染和气候变暖等问题已然成为威胁人类生存与发展的重大危机,如何从根本上解决这些阻碍人类社会可持续发展的障碍,成为包括中国在内的世界各国科学家所共同面临的挑战。其中,许多科学家都把目光投向半导体光催化领域[1-2],这是由于一方面光催化材料具有还原作用,可以利用充足的太阳光催化分解水制氢,提供清洁动力能源,还可以将二氧化碳还原为有机低碳烷烃燃料,降低大气中温室气体的含量;另一方面它还具有氧化作用,可以降解和矿化环境中的各种有机、无机污染物,解决环境污染等问题。以二氧化钛(TiO_2)和氧化锌(ZnO)等为代表的传统光催化材料因具有较宽的带隙,只能利用太阳光中的紫外光($<400nm$),极大地限制了它们对太阳能的利用效率。最近十几年来,具有可见光吸收特性的半导体光催化纳米材料引起了人们的广泛关注[3-5]。

铋系光催化材料在有效处理含有有机物的废水以及空气中的有机物等方面具有很大的发展潜力,渐渐地受到人们的重视。相对而言,催化剂表面催化活性位点增加以及对可见光的强吸收,使铋系光催化材料成为研究的热点。而对于影响光催化效率的一些内在因素(比如:探究其内部构效关系,抑制它的复合概率,并且相应提高其氧化能力,从而得到高效、稳定的可见光催化剂),成为了今后研究的大致方向。其中,为了实现实际生活中高效地利用光催化剂处理环境污染问题,需要对光催化过程以及机理有充分的了解,以提高对光催化技术的认识程度。光催化是光和催化剂共同作用而引发的氧化还原反应。由于半导体能带的不连续,在填满电子的低能价带和空的高能导带之间存在一个禁带,所以当用能量等于或大于禁带宽度(一般在3eV以下)的光照射半导体时,其价带上的电子被激发,越过禁带进入导带,同时在价带上产生相应的空穴。在半导体水悬浮液中,在能量的作用下电子和空穴分离并迁移到粒子表面的不同位置参与氧化还原反应,最终生成氧化能力极强的羟基自由基。

四方晶系的BiOI禁带宽度为1.77eV,其化学稳定性和光催化性能优异引起了研究者的极大兴趣[6-10]。BiOI虽然具有很强的可见光响应能力,但其自身载流子复合率高、光化学反应活性低,限制了其应用[11]。研究表明光催化剂与其他材料复合,能促进光生载流子的有效分离,提高其光催化效率[12-20]。

MOFs 材料具有结构奇特、化学性质多样、比表面积大、孔径可调等优点，广泛应用于催化、分子吸附和分离以及多孔载体等领域[21-26]。MOFs 材料具有吸收可见光的能力，是制备光催化剂的理想材料[27-30]。

作为 MOFs 中的一员，UiO-66 由于具有高比表面积、发达的微孔结构和优异的结构稳定性，在吸附方面具有潜在的应用前景，因此引起了大量的研究。UiO-66 的结构有两种形式：羟基化和脱羟基化。在室温下存在的 UiO-66 结构富含—OH，并且当其在 200～330℃经历热处理时，会经历脱羟基化过程，来自 Zr-O 金属簇的—OH 以 H_2O 形式离开并且 Zr 从八配位形式转变为七配位形式以形成脱羟基的 UiO-66。有人研究了羟基化和脱羟基结构对 UiO-66 气体吸附性能的影响：首先，通过对脱羟基的 UiO-66 在 25℃和 80℃下进行水蒸气吸附连接循环测试，UiO-66 的羟基化和脱羟基化在一定条件下可以相互转化，并保持结构稳定；其次，结果表明，羟基化和脱羟基化的结构差异对 UiO-66 的吸附能力没有显著影响，在不同压力下，对 UiO-66 的吸附性能也没有显著影响。对于对极性气体（如 CO_2）的吸附 UiO-66 表现出与 NaY 沸石相当的吸附能力，其吸附能力主要由微孔骨架结构决定，而不是与 CO_2 的化学配位，这使得 UiO-66 的重现性较好。可以看出，具有高稳定性和高孔隙率的基于 UiO-66 的材料可成为高效 CO_2 吸附剂之一。采用合成后交换法用 Ti 代替 UiO-66 中的一些 Zr，形成 Zr-Ti 双金属 MOFs 以吸附 CO_2，结果表明：在较小的原子半径的 Ti 代替 Zr 之后，形成具有比结构中的 Zr-O 键长小的 Ti-O 键，导致正八面体的孔径减小，孔径的减小不仅可以使孔径接近 CO_2 的理想吸附孔径，而且还可以增加 CO_2 吸附的等效热量，使更多的有效势能在吸附表面之间传递，从而提高吸附容量。0℃时，56%Ti 取代的 UiO-66 的吸附容量为 4mmol/g，比未取代的 UiO-66 的吸附容量高 81%。官能化的 UiO-66-X 对 CO_2 的吸附由于配体不同也显示出一些差异。首先在 0℃和 20℃、0.1MPa 下研究了 UiO-66-$(CH_3)_2$、UiO-66-NH_2、UiO-66-NO_2 和 UiO-66-Br 的吸附能力。—$(CH_3)_2$—NH_2 和—NO_2 基团的引入可以不同程度地提高对 CO_2 的吸附容量，其中 UiO-66-$(CH_3)_2$ 具有最高的吸附容量，与 UiO-66 相比，它增加了 33%，而—Br 则表现出相反的效果。UiO-66-$(CH_3)_2$ 具有最高的 CO_2 吸附容量，因为甲基的引入适当地降低了 UiO-66 的孔径尺寸，提高了 CO_2 的吸附容量[15]。研究表明，UiO-66 和 UiO-66-NH_2 的活化过程对催化活性具有非常重要的影响。

在 150℃ 空气中活化的 UiO-66 作为催化剂，反应 1h 后，反应物的转化率仅为 30%。在 300℃ 下真空活化后，反应物的转化率可增加到 42%。活化后 UiO-66 反应选择性约为 80%。UiO-66-NH_2 催化剂表现出较高的催化活性，在 150℃ 空气中活化的 UiO-66-NH_2 作为催化剂，反应物的转化率为 67%，而真空活化后，反应物的转化率仅为 38%，用两种活化方法处理的 UiO-66-NH_2 反应选择性为 90%，UiO-66 和 UiO-66-NH_2 可以通过以下事实来解释，高温活化可导致 UiO-66 脱去羟基，其结构中更多的配位不饱和 Zr 位点导致酸性和催化活性增加。然而，完全脱羟基化的 UiO-66 中过强的酸性位点也会导致目标产物的反应选择性降低，具有碱性—NH_2 基团的 UiO-66-NH_2 是具有酸性和反应性位点的催化剂[16]，位点可以活化苯甲醛，而碱性—NH_2 基团可以活化脂肪醛，从而使 UiO-66-NH_2 具有更好的催化活性。

使用—SO_3H 官能化的 UiO-66-SO_3H 作为傅里德-克拉夫茨反应用苯甲酰氯对二甲苯进行手性酰化的催化剂，还发现了 UiO-66 结构的酸性和催化活性有直接的关系。他们分别通过原位合成和合成后交换制备了 UiO-66-SO_3H。通过 NH_3-TPD（程序升温脱附）发现 UiO-66 中存在更多的弱酸性和中等酸性位点，并且 UiO-66-SO_3H 具有更多的强酸性位点，原位合成的 UiO-66-SO_3H 的总酸量最高[17]。每种催化剂的反应结果表明，具有最高总酸量的 UiO-66-SO_3H 具有最高的苯甲酰氯转化率，而 UiO-66 具有最低的转化率。UiO-66 作为酸催化剂，其催化效率仍然低于传统的酸催化剂，UiO-66 和其他 Zr 基 UiO 系列 MOFs 材料中的 [$Zr_6O_4(OH_4)$] 金属团簇的配体数量最多，使该物质成为 MOFs 家族稳定的一个整体。UiO-66 或配体功能化的 UiO-66-X 在不同研究领域具有较高的研究价值，并且具有一定的应用前景。与传统的多孔材料如分子筛相比，UiO-66 还存在耐碱性差、收率低等缺点，但迄今为止，对 UiO-66 或其他 MOFs 材料的结构优化和性能开发的研究尚处于起步阶段。我们认为，在不久的将来，对 MOFs 的学术研究和产业化将取得更丰富、更有价值的成果，实际应用保证了源源不断的动力和支持。

UiO-66 的固体样品在一般情况下都是通过经典的溶液热法获得的。为了生成结晶母液将 $ZrCl_4$ 和配体 1,4-对苯二甲酸（H_2BDC）溶解在 DMF 溶剂中并在 120℃ 下形成静态晶体。然后在 24h 后进行洗涤和干燥，得到一个晶体大小约为 150nm、大致为立方体的 UiO-66 样品。在合成 UiO-66 的过程中有一个重要的影响因素，为了实现 UiO-66 的受控合成，通常会添加不同类型的调节剂以便改变结晶系统和配位环境的反应速率[11]。在 UiO-66 的结晶体中，H_2BDC 会与具有羧基的改性调节剂形成 $[Zr_6O_4(OH)_4]$ 金属簇的配位配体。当然竞争协调会在一定程度上损害配体和金属簇之间的配位平衡，因此 UiO-66 的合成通常会受到抑制，不仅速率降低而且会有小的核颗粒产生。这时如果用氢氟酸作调节剂，会因为氟离子的强电负性，快速成为 H_2BDC 的一部分，从而与 $[Zr_6O_4(OH)_4]$ 配位且取代氯离子而补充 UiO-66 结构中的配体缺陷，以便于维持骨架的电荷平衡，氟离子的引入会在一定程度上阻止 H_2BDC 与 $[Zr_6O_4(OH)_4]$ 的反应，从而影响晶体成核和生长的反应速率以及实现 UiO-66 减小晶体尺寸和形貌的控制。

多孔结晶材料的晶体尺寸和形貌的变化常常引起材料本身的物理或化学性质的变化。羧酸调节剂除了对结晶速率的影响之外，它们还参与了 $[Zr_6O_4(OH)_4]$ 调节基团形成配位点，在 Zr 中心，这些 Zr 配位修饰基团被加热活化时分解，导致结构中存在更多的配位缺陷位点，配位体缺陷位点越多，孔隙率就越大[12]。

与传统的多孔材料相比，MOFs 的独特之处在于通过选择具有官能团的配体，原位合成或合成后交换功能化的结构使其化学官能化，允许 MOFs 在结构上具有特定的官能团或金属离子，从而有目的地合成具有独特功能的 MOFs。使用具有特定官能团的 H_2BDC-X 而不是 H_2BDC 作为配体（如 H_2BDC-NH_2、H_2BDC-Br、H_2BDC-NO_2）原位合成 UiO-66-X，通过引入官能团来改变 UiO-66 的物理或化学性质，功能化的 UiO-66-X 保持与 UiO-66 具有相同的晶体结构，但由于官能团的不同，比表面积在 1500～6000m^2/g 之间变化。引入具有较大体积和质量的官能团，例如—Br 和—NO_2，减少了 UiO-66 结构中孔隙的自由空间，导致 UiO-66-Br 和 UiO-66-NO_2 的比表面积与 UiO-66 相比更小，UiO-66-NH_2 孔隙空间受—NH_2 影响较小，因此其比表面积与 UiO-66 相当。UiO-66-Br 类似于 UiO-66，其结构在 450℃稳定，并且 UiO-66-NH_2 和 UiO-66-NO_2 的稳定性相对较低，并且在 350℃开始溶解。然而，某些官能团不能通过原位合成与 UiO-66 的结构相连，因此研究人员着重于使用简单而有效的合成后交换法（PSE）获得 UiO-66-X。在溶液中用 2，3-二羟基对苯二甲酸（CAT-BDC）进行了 UiO-66 的配体交换。由于 CAT-BDC 与 H_2BDC-

BDC（UiO-66 结构中的一部分 H_2BDC）之间具有大的结构相似性，因此—OH 基团在 UiO-66 上分离，这使 UiO-66-CAT 保留了 UiO-66 的较高结晶度。

在各种 MOFs 材料中，UiO-66 是含锆 MOFs 材料，有较高的热稳定性和化学稳定性，以及与其他 MOFs 材料一样的优异性能。此外，UiO-66 显示出 n 型半导体特征，并在水中表现出优异的结构稳定性[31-32]。因此，UiO-66 是一种很有前途的光催化剂，用于水介质当中，如水处理和水分解制氢制氧。然而，据我们所知，迄今为止关于用于水处理的 UiO-66 基光催化剂的报道很少，将 p 型 BiOI 与 n 型 UiO-66 复合作为用于水处理的光催化剂还没有被报道[29-33]。

基于以上思路，本实验将 BiOI 光催化剂和 UiO-66 进行复合，着力设计制备 UiO-66/ 新型碘氧化铋（UiO-66/BiOI）复合光催化材料，研究其在可见光下的光催化降解率以及光电转换效率。探索出一条制备高活性复合光催化材料的有效途径，从而为高效光催化剂的制备提供理论依据和材料保障。

本实验用低温共沉淀法将 BiOI 纳米粒子均匀负载在 UiO-66 上，制备出了一系列的 UiO-66/BiOI 光催化剂。对所制备的复合光催化剂使用 X 射线衍射、扫描电镜、透射电镜、比表面积测试和紫外 - 可见漫反射光谱进行测试分析，得到了它的微观结构和光学性质的特征。测定其 COD 和 TOC，来确定降解是否无毒环保。通过复合光催化剂在可见光下降解 H 酸，来评估它的光催化活性[21]。最后，从连续五个磁力处理周期的溶液中，分离制备样品。筛选出具有最佳可见光吸收效率的复合光催化材料。研究 UiO-66/BiOI 复合光催化材料的组成、结构和光催化反应活性以及降解 H 酸的构效关系，从微观上认识此类复合光催化材料和光催化反应的本质，从而提高异质纳米光催化材料的量子效率及反应活性。

4.2 实验部分

4.2.1 材料制备

UiO-66/BiOI 复合材料通过简单水热 - 共沉淀法制备。通常情况下，将 0.3mmol $Bi(NO_3)_3 \cdot 5H_2O$ 在剧烈搅拌下溶于 40mL 去离子水中，用 0.5mol/L 硝酸（HNO_3）调节 pH，使 pH=2.0，得到透明的溶液 A。同时，将 UiO-66 和 0.3mmol KI 分散在 30mL 去离子水中并超声分散以获得均匀分散体 B。然后，在搅拌下将分散体 B 滴加到溶液 A 中。随后，通过滴加 2.0mol/L $NH_4 \cdot H_2O$ 最终将悬浮液的 pH 值调节至 3.0，并转移至 100mL 反应釜中，并在连续搅拌下保持在 85℃ 24h。收集形成的红色沉淀并分别用去离子水和乙醇洗涤数次，然后将样品在室温下在真空烘箱中干燥 24h。为了比较，纯 BiOI 纳米片也通过相同的程序制备而不引入 UiO-66。该实验的主要步骤及技术路线如图 4-1 所示。

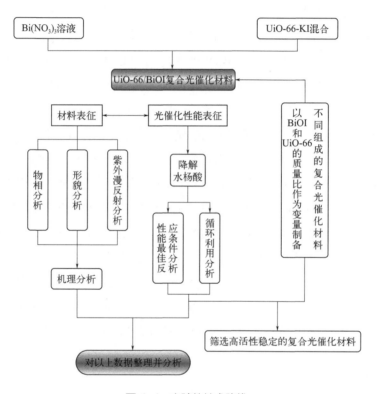

图 4-1 实验的技术路线

4.2.2 实验取样分析

在实验装置中，使用光强度为 $300mW \cdot cm^{-2}$ 的氙气灯作为光源，并使用 420nm 截止滤光片（中国徐江机电厂）且仅提供可见光照射。将约 100mg 光催化剂加入 100mL 水杨酸溶液（c=10mg/L）中。在照射之前，将悬浮液在黑暗中磁力搅拌 3h 以达到光催化剂和水杨酸之间的吸附-解吸平衡，然后在磁力搅拌下将溶液暴露于可见光照射下。从悬浮液中收集约 3mL 等分试样，并且每经过 10 分钟立即离心。使用 UV-2501PC 分光计测定 296nm 处的吸光度，以此来监测水杨酸的降解。所有的光催化剂实验均执行 5 次并具有重现性。计算后的图形结果表示为平均值和标准偏差。

4.2.3 表征

样品的粉末 X 射线衍射（XRD）图像通过具有 CuKα 射线（$\lambda=0.15406nm$）的布鲁克 AXS D8 Advance 粉末衍射仪记录。在 JEOL HitachiS-4800 场发射扫描电子显微镜（FESEM）上观察合成样品的形态。在 JEOL JEM-2100 显微镜上拍摄高分辨透射电子显微镜（HRTEM）图像。在具有 $BaSO_4$ 作为参考的紫外 - 可见光谱仪上进行样品的紫外 - 可见漫散射光谱。使用 BET 方法计算样品的表面积。

4.3 结果与讨论

采用简单的水热共沉淀法合成了一系列 UiO-66 含量不同的 UiO-66/BiOI 复合光催化材料。首先 UiO-66 通过典型的水热法制备，然后在 BiOI 的前驱体溶液中加入 30、60 和 90mg 的 UiO-66，样品命名为 UiO-66/BiOI-1、UiO-66/BiOI-2 和 UiO-66/BiOI-3，其中 UiO-66 的质量分数分别为 2.6%、5.2% 和 7.8%。纯 BiOI、UiO-66 和含不同含量 UiO-66 的 UiO-66/BiOI 复合光催化材料的 XRD 图谱如图 4-2 所示。所合成的 UiO-66 的 XRD 图谱可以与先前的报道

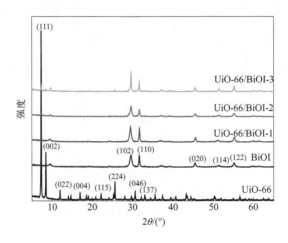

图 4-2　BiOI、UiO-66 和 UiO-66 含量不同的 UiO-66/BiOI 复合材料的 XRD 图谱

进行对比，表明 UiO-66 已成功合成。在 UiO-66/BiOI 复合光催化材料的 XRD 图谱中观察到 BiOI 和 UiO-66 的特征峰，并且随着 UiO-66 含量的增加，UiO-66 的特征峰变强。

通过 SEM、TEM 和 HRTEM 观察了 UiO-66、BiOI 和 UiO-66/BiOI-2 的形貌和微观结构。如图 4-3（a）所示，UiO-66 是边长平均为 500nm 的立方体。图 4-3（b）显示 BiOI 由不规则的光滑片构成，并且这些片的尺寸约为 1μm。图 4-3(c) 和图 4-3(d) 是 UiO-66/BiOI-2 复合材料的 SEM 和 TEM 图像，显然，BiOI 片材表面上附着着 UiO-66 立方体。如图 4-3（e）所示，附着的不规则平滑片具有 0.298nm 的晶格条纹间隔，对应于四方 BiOI 的（102）晶面。然而 UiO-66 的 HRTEM 图像不能获得，因为在高能电子束照射下 UiO-66 微晶倾向于被损坏。

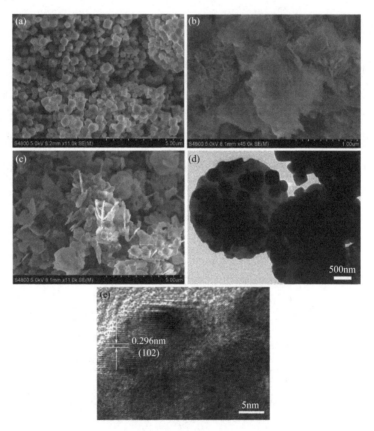

图 4-3　UiO-66（a）、BiOI（b）和 UiO-66/BiOI-2（c）的 SEM 图像；UiO-66 / BiOI-2 的 TEM（d）和 HRTEM（e）图像

使用 N_2 吸附-解吸测量并研究样品的 BET 比表面积。如表 4-1 所示，UiO-66 和 BiOI 的 BET 比表面积分别为 527.4072$m^2 \cdot g^{-1}$ 和 16.1617$m^2 \cdot g^{-1}$。值得注意的是，UiO-66 的含量极大地影响 UiO-66/BiOI 复合材料的 BET 比表面积。与纯 BiOI 相比，UiO-66/BiOI 复合材料的 BET 比表面积随着 UiO-66 含量的增加从 68.4469$m^2 \cdot g^{-1}$ 逐渐增加到 300.59572$m^2 \cdot g^{-1}$。多孔 UiO-66 具有较大的比表面积，因此随着 UiO-66 含量的增加，UiO-66/BiOI 复合材料的 BET 比表面积逐渐增大。较大比表面积的光催化剂可以提供更多的表面活性位点并促进电荷载体转移，从而提高光催化性能。

表 4-1 UiO-66、BiOI 和 UiO-66 含量不同的 UiO-66/BiOI 的多孔结构参数

样品	比表面积/($m^2 \cdot g^{-1}$)	总孔容/($cm^3 \cdot g^{-1}$)
UiO-66	527.4072	0.280551
BiOI	16.1617	0.124270
UiO-66/BiOI-1	68.4469	0.050094
UiO-66/BiOI-2	138.7424	0.117471
UiO-66/BiOI-3	300.5957	0.176943

通过紫外-可见光谱仪测试了制备的样品在不同波长下的吸收能力（图 4-4）。可以看出，UiO-66 的基本吸收边约在 320nm 处，表明 UiO-66 不具有吸收可见光的能力。纯 BiOI 的吸收边约在 620nm 处，在可见光范围内具有

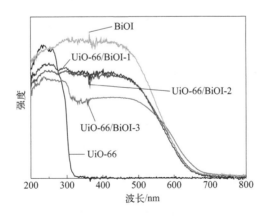

图 4-4 UiO-66、BiOI 和 UiO-66 含量不同的 UiO-66/BiOI 复合光催化材料的紫外-可见漫反射光谱

强吸收。具有不同 UiO-66 含量的 UiO-66/BiOI 复合光催化材料表现出纯 UiO-66 和 BiOI 混合吸收的特征。UiO-66/BiOI-1、UiO-66/BiOI-2 和 UiO-66/BiOI-3 的吸收特征为随着 UiO-66 含量的增加，UiO-66 的吸收边更加明显。

  BiOI、UiO-66 和 UiO-66/BiOI 复合材料的光催化活性通过在可见光照射下降解水杨酸来评估。如图 4-5 所示，在没有光催化剂的情况下，水杨酸自身光降解在 120min 内几乎观察不到，这表明在入射光下水杨酸非常稳定。随着 UiO-66 含量的增加，没有可见光照射的情况下水杨酸浓度逐渐下降，这是由于复合光催化剂的吸附作用。多孔 UiO-66 具有较大的比表面积，可以增强对水杨酸的吸附。适量地添加 UiO-66 可以有效地增大其比表面积以及提高 UiO-66/BiOI 纳米复合材料的吸附能力。然而，随着 UiO-66 的含量进一步增加，由于 UiO-66 的过量覆盖，BiOI 纳米颗粒的有效反应表面会减少，由此导致可见光照射下光催化活性的快速下降。因此，与 BiOI 复合时，只有当 UiO-66 的质量分数适当（5.2%），UiO-66/BiOI-2 纳米复合材料的光催化率才能达到最大值。所有 UiO-66/BiOI 复合光催化材料均可在 120min 内将大部分水杨酸分子降解，而 UiO-66/BiOI-2 样品的降解率最高。相比之下，纯 BiOI 降解水杨酸率比 UiO-66/BiOI 复合材料低。

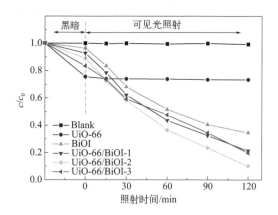

图 4-5 可见光照射下不同光催化剂对水杨酸的光催化降解

  光催化剂的稳定性对于它们的实际应用效果非常重要。因此研究了 UiO-66/BiOI 复合光催化材料在可见光照射下光催化降解水杨酸的稳定性。每次光降解后，通过离心机将光催化剂从溶液中分离，并在用去离子水超声波清洗后重新使用，其他因素保持一致。如图 4-6 所示，虽然每个循环的光催化率略有下降，但在第 5 个循环中仍有 85% 以上的水杨酸能被降解，这表明 UiO-66/BiOI 复合光催化材料在可见光下的光催化反应中具有良好的稳定性。

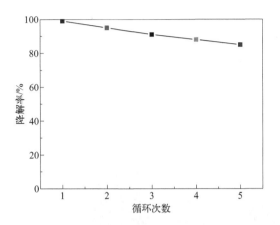

图 4-6　UiO-66/BiOI 复合材料用于水杨酸光催化降解（光催化剂用量为 0.1g·L^{-1}，水杨酸浓度为 10mg·L^{-1}，pH 值为 7，照射时间为 120min）的可重复使用性

根据上述结果和以前的一些文献[15, 33-34]，UiO-66/BiOI 光催化活性的增强可以归因于以下几方面。首先，对于光催化降解过程，反应物的初步吸附对于有效降解非常重要。与纯 BiOI 相比，UiO-66/BiOI 复合材料具有相对较大的比表面积，因此，UiO-66/BiOI 复合材料可以在其表面吸附更多的反应物，从而更容易地发生氧化还原反应。其次，UiO-66/BiOI 复合材料中 UiO-66/BiOI p-n 结的形成在光催化降解水杨酸中起着重要作用，因此我们提出了一种可能的机理。如图 4-7 所示，BiOI 是 p 型半导体，根据之前的报道，BiOI 的价带（VB）顶部和导带（CB）底部分别为 2.30 和 0.58eV[35]。UiO-66 是一种 n 型半导体，VB 和 CB 的边缘位置分别为 2.9 和 -0.6eV[36]。当 BiOI 和 UiO-66 接触并暴露于可见光照射下时，UiO-66 和 BiOI 都可以被激活。UiO-66 的 VB 中的电子被激发到潜在边缘（-0.60eV），而 BiOI 的 VB 中的那些可以激发至更高的电势边缘（-0.68eV）[37]。BiOI 的重整 CB 电位（-0.68eV），比 UiO-66（-0.60eV）更负，BiOI 的 VB 电位（2.30eV）比 UiO-66（2.90eV）的阳性小。所以 UiO-66/BiOI p-n 结属于 B 型异质结[38]，因此，BiOI 的 CB 上的光致电子可以轻松地迁移到 UiO-66 的 CB 上，因为 UiO-66 的 VB 上的光生空穴将转移到 BiOI 的 VB 上。由 BiOI 和 UiO-66 之间的 B 型异质结在界面处产生的电场促进电

子从 p 型半导体的 VB 迁移到 n 型半导体的 VB，这将大大降低光生电子和空穴的重组率。这些孔可以直接氧化有机分子。UiO-66/BiOI 的新 CB 电位比 $E(O_2/\cdot O_2^-)$（-0.33eV）更负 [39]，因此 UiO-66/BiOI 新 CB 中的光生电子可与 O_2 反应形成 $\cdot O_2^-$，然后与有机分子反应使其降解。UiO-66/BiOI 复合材料比纯 BiOI 和 UiO-66 具有更高的光催化活性。

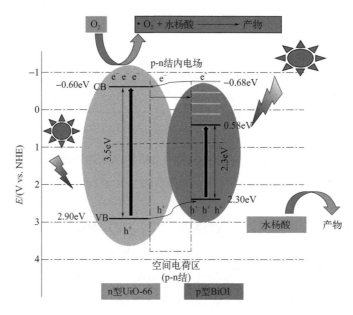

图 4-7　BiOI、UiO-66 的能量位置和 p-n 结的形成以及可能的电荷分离过程的示意图

4.4　结论

综上所述，通过简便的水热法制备了一系列具有不同 UiO-66 含量的 UiO-66/BiOI 异质结光催化剂。当 UiO-66 质量分数为 5.2% 时，所得的 UiO-66/BiOI 复合材料对可见光照射下水杨酸的降解具有最佳的光催化活性。此外，我们提出了 UiO-66/BiOI 异质结光催化剂的光催化机理。UiO-66/BiOI 复合材料的光催化活性增强可能是 BiOI 与 UiO-66 之间形成 p-n 结的结果，这可以促进光激发电子 - 空穴对的分离。UiO-66/BiOI 复合材料在回收实验中表现出高的光稳定性。这些结果表明获得的 UiO-66/BiOI 复合材料是潜在的可见光光催化材料。

参考文献

[1] Shang L, Tong B, Yu H, et al. CdS nanoparticle-decorated Cd nanosheets for efficient visible light-driven photocatalytic hydrogen evolution[J]. Adv Energy Mater, 2016, 6: 1501241-1501247.

[2] Jiao Z, Zheng J, Feng C, et al. Co-doped $BiVO_4$ photoanodes with a metal-organic framework cocatalyst for improved photoelectrochemical stability and activity[J]. ChemSusChem, 2016, 9: 1-9.

[3] Zhou Q, Peng F, Ni Y, et al. Long afterglow phosphor driven round-the-clock $g-C_3N_4$ photocatalyst[J]. Photochem Photobiol A 2016, 328: 182-188.

[4] Zheng J, Jiao Z. Modified Bi_2WO_6 with metal-organic frameworks for enhanced photocatalytic activity under visible light[J]. Colloid Interface Sci, 2017, 488 : 234-239.

[5] He Z, Shi Y, Gao C, et al. $BiOCl/BiVO_4$ p-n heterojunction with enhanced photocatalytic activity under visible-light irradiation[J]. J Phys Chem C, 2013, 118: 389-398.

[6] Cheng H, Huang B, Dai Y. Engineering BiOX (X = Cl, Br, I) nanostructures for highly efficient photocatalytic applications[J]. Nanoscale, 2014, 6: 2009-2026.

[7] Cao S, Zhou P, Yu J. Recent advances in visible light Bi-based photocatalysts[J]. Chin J Catal, 2014, 35: 989-1007.

[8] He R, Cao S, Guo D, et al. 3D BiOI-GO composite with enhanced photocatalytic performance for phenol degradation under visible-light[J]. Ceram Int, 2015, 41: 3511-3517.

[9] Han S, Li J, Yang K, et al. Fabrication of a β-Bi_2O_3/BiOI heterojunction and its efficient photocatalysis for organic dye removal[J]. Chin J Catal, 2015, 36: 2119-2126.

[10] Mehraj O, Pirzada B, Mir N, et al. A highly efficient visible-light-driven novel p-n junction Fe_2O_3/BiOI photocatalyst: Surface decoration of BiOI nanosheets with Fe_2O_3 nanoparticles[J]. Appl Surf Sci, 2016, 387: 642-651.

[11] Xiao X, Zhang W. Facile synthesis of nanostructured BiOI microspheres with high visible light-induced photocatalytic activity[J]. J Mater Chem, 2010, 20: 5866-5870.

[12] Kim H , Borse P, Choi W, et al. Photocatalytic nanodiodes for visible-light photocatalysis[J]. Angew Chem Int Ed, 2005,117: 4661-4665.

[13] Chang X, Wang T, Zhang P, et al. Enhanced surface reaction kinetics and charge separation of

p-n heterojunction $Co_3O_4/BiVO_4$ photoanodes[J]. J Am Chem Soc, 2015, 137: 8356-8359.

[14] Zhong M, Hisatomi T, Kuang Y, et al. Surface modification of CoO_x loaded $BiVO_4$ photoanodes with ultrathin p-type NiO layers for improved solar water oxidation[J]. J Am Chem Soc, 2015, 137: 5053-5060.

[15] Ye K, Chai Z, Gu J, et al. $BiOI-BiVO_4$ photoanodes with significantly improved solar water splitting capability: P-n junction to expand solar adsorption range and facilitate charge carrier dynamics[J]. Nano Energy, 2015, 18: 222-231.

[16] Xiang Y, Ju P, Wang Y, et al. Chemical etching preparation of the $Bi_2WO_6/BiOI$ p-n heterojunction with enhanced photocatalytic antifouling activity under visible light irradiation[J]. Chem Eng J, 2016, 288: 264-275.

[17] Dai G, Yu J, Liu G. Synthesis and enhanced visible-light photoelectrocatalytic activity of p-n junction $BiOI/TiO_2$ nanotube arrays[J]. J Phys Chem C, 2011, 115: 7339-7346.

[18] Jiang J, Zhang X, Sun P, et al. ZnO/BiOI heterostructures: Photoinduced charge-transfer property and enhanced visible-light photocatalytic activity[J]. J Phys Chem C, 2011, 115: 20555-20564.

[19] Huang H, Xiao K, He Y, et al. In situ assembly of $BiOI@Bi_{12}O_{17}Cl_2$ p-n junction: Charge induced unique front-lateral surfaces coupling heterostructure with high exposure of BiOI {001} active facets for robust and nonselective photocatalysis[J]. Appl Catal B, 2016, 199: 75-86.

[20] Hou D, Hu X, Hu P, et al. $Bi_4Ti_3O_{12}$ nanofibers-BiOI nanosheets p-n junction: Facile synthesis and enhanced visible-light photocatalytic activity[J]. Nanoscale, 2013, 5: 9764-9772.

[21] Ai L, Zhang C, Li L, et al. Iron terephthalate metal-organic framework: Revealing the effective activation of hydrogen peroxide for the degradation of organic dye under visible light irradiation[J]. Appl Catal B, 2014, 148: 191-200.

[22] Horiuchi Y, Toyao T, Saito M, et al. Visible-light-promoted photocatalytic hydrogen production by using an amino-functionalized Ti(Ⅳ) metal-organic framework[J]. J Phys Chem C, 2012, 116: 20848-20853.

[23] Gao J, Miao J, Li P, et al. A p-type Ti(Ⅳ)-based metal-organic framework with visible-light photo-response[J]. Chem Commun, 2014, 50: 3786-3788.

[24] Peng R, Wu C, Baltrusaitis J, et al. Ultra-stable CdS incorporated Ti-MCM-48 mesoporous

materials for efficient photocatalytic decomposition ofwater under visible light illumination[J]. Chem Commun, 2013, 49: 3221-3223.

[25] He J, Yan Z, Wang J, et al. Significantly enhanced photocatalytic hydrogen evolution under visible light over CdS embedded on metal-organic frameworks[J]. Chem Commun, 2013, 49: 6761-6763.

[26] Shen L, Liang S, Wu W, et al. CdS-decorated UiO-66(NH_2) nanocomposites fabricated by a facile photodeposition process: An efficient and stable visible-light-driven photocatalyst for selectiveoxidation of alcohols[J]. J Mater Chem A, 2013, 1: 11473-11482.

[27] Liang Q, Zhang M, Zhang Z, et al. Zinc phthalocyanine coupled with UIO-66 (NH_2) via a facile condensation process for enhanced visible-light-driven photocatalysis[J]. J Alloys Compd, 2017, 690: 123-130.

[28] Shen L, Luo M, Liu Y, et al. Noble-metal-free MoS_2 co-catalyst decorated UiO-66/CdS hybrids for efficient photocatalytic H_2 production[J]. Appl Catal B, 2015, 166: 445-453.

[29] Li S, Wang X, He Q, et al. Synergistic effects in N-$K_2Ti_4O_9$/UiO-66-NH_2 composites and their photocatalysis degradation of cationic dyes[J]. Chin J Catal, 2016, 37: 367-377.

[30] Wang F, Zhang Y, Xu Y, et al. Enhanced photodegradation of Rhodamine B by coupling direct solid-state Z-scheme N-$K_2Ti_4O_9$/g-C_3N_4 heterojunction with high adsorption capacity of UiO-66[J]. J Environm Chem Eng, 2016, 4: 3364-3373.

[31] Yuan Y, Yin L, Cao S, et al. Improving photocatalytic hydrogen production of metal-organic framework UiO-66 octahedrons by dye-sensitization[J]. Appl Catal B, 2015, 168: 572-576.

[32] Luo W, Li Z, Yu T, et al. Effects of surface electrochemical pretreatment on the photoelectrochemical performance of Mo-doped $BiVO_4$[J]. J Phys Chem C, 2012, 116: 5076-5081.

[33] Sha Z, Chan H, Wu J. Ag_2CO_3/UiO-66(Zr) composite with enhanced visible-light promoted photocatalytic activity for dye degradation[J]. J Hazard Mater, 2015, 299: 132-140.

[34] Cavka J, Jakobsen S, Olsbye U, et al. A new zirconium inorganic building brick forming metal organic frameworks with exceptional stability[J]. J Am Chem Soc, 2008, 130: 13850-13851.

[35] Li X, Niu C, Huang D, et al. Preparation of magnetically separable Fe_3O_4/BiOI nanocomposites and its visible photocatalytic activity[J]. Appl Surf Sci, 2013, 286: 40-46.

[36] Zhou J, Wang R, Liu X, et al. In situ growth of CdS nanoparticles on UiO-66 metal-organic

framework octahedrons for enhanced photocatalytic hydrogen production under visible light irradiation[J]. Appl Surf Sci, 2015, 346: 278-283.

[37] Liu Z, Ran H, Wu B, et al. Synthesis and characterization of BiOI/BiOBr heterostructure films with enhanced visible light photocatalytic activity[J]. Colloids Surfaces A, 2014, 452: 109-114.

[38] Shamaila S, Sajjad A, Chen F, et al. WO_3/BiOCl, a novel heterojunction as visible light photocatalyst[J]. J Colloid Interface Sci, 2011, 356: 465-472.

[39] Mousavi M, Habibi-Yangjeh A. Magnetically separable ternary g-C_3N_4/Fe_3O_4/BiOI nanocomposites: Novel visible-light-driven photocatalysts based on graphitic carbon nitride[J]. J Colloid Interface Sci, 2016, 465: 83-92.

第五章

Fe/W 共掺杂 BiVO$_4$/MIL-100（Fe）复合光催化材料制备及性能研究

5.1	引言	130
5.2	实验部分	132
5.3	结果与讨论	133
5.4	结论	142
	参考文献	143

5.1 引言

自从首次在太阳光下使用光电化学（PEC）电池进行水分解以来，人们已经在各种半导体光电极的设计和制造方面做出了巨大的努力[1]。TiO_2是第一个也是研究最广泛地用于PEC实验的阳极。然而TiO_2对可见光的利用不佳以及光生电子-空穴对的高复合率严重限制了其实际应用[2-8]，因此激发了人们在PEC领域中对具有可见光活性和稳定性光电极的探索和开发的兴趣[9-10]。

单斜钒酸铋（m-$BiVO_4$）被确定为一种出色的可见光响应型光电极，可用在PEC水分解中，而且有其他很多优点，例如中等的带隙（$E_g \approx 2.4eV$）、无毒和化学稳定性[11-15]。$BiVO_4$的价带位置合适，位于水氧化电位以下，使其可用于光催化氧的释放[16-18]。尽管$BiVO_4$的导带不为负，但在没有施加偏压足以减少H^+的情况下，它仍然是有希望用于水分解的光电极[19-25]，其中重要的原因之一是$BiVO_4$的光电流阈值是0.0V（标准的"零电位"），这是因为它的导带非常接近氢的激发态。因此，在低偏置区域中，它比具有更多正导带的光电极表现出明显更大的光电流。然而，由于电荷分离和传输特性不足以及表面吸收能力差，纯$BiVO_4$通常表现出相对较低的光催化性能[26-29]。因此，人们采取了许多策略来突破这些局限，包括异质结构的构建[13, 23, 30]、助催化剂的负载[22, 24, 25, 31]和杂质掺杂[12, 16, 17, 26]。

目前已经有许多关于改善$BiVO_4$的PEC性能方法的报道。其中一个方法是通过n型掺杂方法改善$BiVO_4$的PEC性能，即一小部分的V^{5+}被较高价的离子取代（例如Mo^{6+}或W^{6+}）[25, 29]。据报道，将较高价态的离子掺入$BiVO_4$可以提高载流子密度。通过扩大吸收范围并通过在这些导体-液体界面附近产生电场来改善电荷分离[25]，结果表明可使$BiVO_4$的PEC活性大大提高。然而，$BiVO_4$的另一个大问题仍未解决，即PEC稳定性。虽然$BiVO_4$的化学性质在水溶液中稳定，但这并不表示$BiVO_4$可以在PEC实验期间稳定地以设定值输出光电流，这可能受其晶格缺陷或固有特性的影响。可以很容易地发现在PEC稳定性测试期间，$BiVO_4$的光电流急剧下降。尽管用较高价的离子取代V离子可以明显提高光电流值，但不能提高PEC的稳定性。因此，$BiVO_4$半导体中仍然存在顽固的问题，如何解决这个问题仍然是一个充满挑战和有意义的项目。

此外，尽管离子掺杂可以通过在半导体-液体界面附近产生电场来改善电荷分离，但是其能力和作用由于不可避免地存在结构缺陷而受到限制，这些缺陷可作为光生载流子的复合中心。因此，有必要从外部引入助催化剂有效地增强光生载流子的分离[19, 25]。贵金属通常用作光催化 H_2 释放或 CO_2 还原的助催化剂。然而，由于它们的高昂价格，大规模应用成本太高。因此，科研工作者相继制备出了许多无贵金属但高效的助催化剂，例如 WC[32]、MoS_2[33]、WS_2[34]、CoPi[25]、CuO[35-37] 等。与这些传统的无机材料相比，金属有机骨架（MOFs）具有较大的孔隙率和比表面积，这使其成为光催化应用中与半导体复合的热门材料[38-39]。而据我们所知，至今为止没有使用 MOFs 作为助催化剂的，因此我们希望制备复合材料研究这种结构是否可以用作改善 PEC 性能的有效助催化剂。

基于以上研究，我们提出了同时提高 $BiVO_4$ PEC 稳定性和光电极活性的新策略。我们通过掺杂 Fe^{3+} 代替 Bi^{3+} 成功地增强了 $BiVO_4$ 的 PEC 稳定性，这与以前有关 V^{5+} 替代的研究完全不同。密度泛函理论（DFT）计算表明，Fe 掺杂会在 $BiVO_4$ 的禁带中产生杂质带，从而缩小能带并扩大光吸收范围。Fe 掺杂的 $BiVO_4$ 可以保持光电流稳定性，这可能是因为用 Fe^{3+} 取代 Bi^{3+} 可以克服 $BiVO_4$ 的结构缺陷。更重要的是，Fe 掺杂不仅有利于 $BiVO_4$ 的稳定性和光转换效率，而且还可以与其他手段协同作用，进一步提高 $BiVO_4$ 的 PEC 活性。结果表明，仅采用 Fe 掺杂，$BiVO_4$ 即可提供高稳定的 PEC 性能，无论是掺杂其他杂质还是负载助催化剂。

因此，在这项工作中，我们首先制备了纳米多孔 Fe/W 共掺杂 $BiVO_4$ 光电极改善其 PEC 稳定性和活性，然后将 MIL-100（Fe）用作助催化剂，以进一步促进电荷载体的分离。在这个新颖的系统中，Fe、W 和 MIL-100（Fe）分别充当稳定剂、n 型掺杂剂和助催化剂。特别是掺杂 Fe^{3+} 代替多孔 $BiVO_4$ 光子中的 Bi^{3+} 位确实可以显著改善 PEC 稳定性并保持稳定的光电流。用 W^{6+} 代替 V^{5+} 是一种典型的 n 型掺杂，可以通过增加载流子密度和促进电荷分离来大大提高 PEC 的性能[29]。此外，发现 MIL-100（Fe）可以用作促进电荷分离的有效助催化剂，从而进一步提高 PEC 活性。实验结果表明，掺杂会使光电极光电流密度增加 7 倍，添加 MIL-100（Fe）作为助催化剂时可以扩大到 14 倍。

5.2 实验部分

5.2.1 纳米多孔 $BiVO_4$ 和 Fe/W 共掺杂 $BiVO_4$ 的制备

通过滴铸法制备了纳米多孔 $BiVO_4$ 光电极。通过以下步骤获得前驱体溶液。第一，将 $Bi(NO_3)_3 \cdot 5H_2O$ 和 NH_4VO_3 分别溶解在 75mL 的乙二醇溶剂中。第二，将 0.68g 聚乙二醇-600（PEG-600）溶解在另一 20mL 乙二醇溶液中。其次，根据化学计量比，通过以下配方将三种溶液混合：5mL $Bi(NO_3)_3 \cdot 5H_2O$ 溶液、5mL NH_4VO_3 溶液和 2.5mL PEG-600 溶液。第三，将 0.2mL 前驱体溶液滴铸在掺杂氟的 SnO_2 导电玻璃（FTO）基底上。最后将样品在 150℃的烘箱中干燥 60min，然后于马弗炉中在 500℃的温度下退火 2.5h。并用相同的方法合成了 Fe/W 共掺杂的 $BiVO_4$。第一步不同之处在于：$Bi(NO_3)_3 \cdot 5H_2O$ 和 $Fe(NO_3)_3$ 溶解在乙二醇溶液中，Fe 的量等于 3%（摩尔浓度）。同样，NH_3VO_3 和 $(NH_4)_{10}H_2(W_2O_7)_6$ 溶解在另一乙二醇溶液中以 3%（摩尔浓度）计。这项研究首次将 MIL-100（Fe）作为助催化剂来提高光电极的光电化学性能，这可以为 PEC 应用提供替代选择。

5.2.2 MIL-100（Fe）纳米粒子的制备

根据文献报道，通过水热法合成了 MIL-100（Fe）[40]。在典型的反应中，H_3BTC（0.8466g）和 $FeCl_3 \cdot 6H_2O$（1.6263g）溶解在蒸馏水（30mL）中，然后加入 HF（0.213mL）和 HNO_3（0.163mL），然后将溶液转移到反应釜中在 150℃下放置 12h。通过过滤回收橙色固体，用热水和乙醇洗涤，最后在真空下 150℃干燥 12h。

5.2.3 Fe/W 共掺杂 $BiVO_4$/MIL-100（Fe）复合光催化材料的制备

首先将 MIL-100（Fe）分散到乙醇溶液中，然后再通过旋涂法将其负载在 Fe/W 共掺杂的 $BiVO_4$ 光电极上。用 MIL-100（Fe）改性的 Fe/W 共掺杂 $BiVO_4$ 在 150℃下干燥 30min。

5.2.4 理论计算

使用从头计算量子力学程序（CASTEP）中实现的第一性原理（GGA-PBE）[41]进行计算[42]。应用超软赝势来描述核心电子与价电子之间的相互作用。Bi、V 和 O 的价电子配置分别为 $6s^26p^3$、$3s^23p^63d^34s^2$ 和 $2s^22p^4$。Fe 和 W 的价电子构型分别为 $3d^64s^2$ 和 $5s^25p^65d^46s^2$。平面波截断能为 600eV，偏聚能为 0.316eV。

5.2.5 表征

XRD 测量是在仪器上使用 CuKα 辐射（40kV）进行的。XRD 扫描范围是从 10° 到 90°，扫描速率为 $0.0678s^{-1}$。在场发射扫描电子显微镜上以 5kW 的加速电压进行 SEM 测量。TEM 测量是通过使用 200kW 的透射电子显微镜进行的。使用 $BaSO_4$ 作为背景，在紫外-可见光谱仪上获得紫外-可见光漫反射光谱。

5.2.6 光电化学性能检测

使用 MIL-100（Fe）改性的 Fe/W 共掺杂 $BiVO_4$ 作为 PEC 电池中的光电极。光电流响应 CHI-660D 稳压器以 Pt 切片作为夹心型配置，在可见光照射下记录。对于电极，用饱和甘汞电极（SCE）作为参比电极，以 0.1mol/L Na_2SO_4 溶液作为电解质。装有模拟太阳滤光片将 300W 氙弧灯，模拟太阳滤光片校准为 $100mW \cdot cm^{-2}$，并使用辐射计作为光源。使用 300W Xe 灯和单色仪在 0.8V 下对 SCE 进行单色光下的光电流响应测量。莫特-斯科蒂图是使用电化学分析仪在标准三电极系统中以 5kHz 的频率进行测试的。设置频率范围为 $0.1 \sim 10^5 Hz$，电压为 0.6V，交流（AC）幅值为 5mV，测试电化学阻抗谱（EIS）奈奎斯特（Nyquist）图。所有实验均在室温条件下进行。

5.3 结果与讨论

多孔 $BiVO_4$ 光电极是通过简便的滴铸法制造的（图 5-1）[43]。为了比较，掺 Fe $BiVO_4$ 和 W 掺杂的 $BiVO_4$ 也用相同的方法制备，除了分别在生长溶液中添加 Fe 源和 W 源。从紫外-可见吸收光谱中[图 5-2（a）]可以看出，Fe 掺杂有效地缩小了 $BiVO_4$ 的带隙，这与下面的理论计算是一致的。$BiVO_4$ 的 PEC 稳定性问题通过线性扫描伏安图[LSV，图 5-2（b）]得以解决。用不同速率扫描时，电流-电位特性存在很大差异。以 0.6V 为例，用 $0.1V \cdot s^{-1}$ 扫描速率扫描时的光电流为 $1.2mA \cdot cm^{-2}$，但是用 $0.01V \cdot s^{-1}$ 的扫描速率扫描时，它减小到 $0.2mA \cdot cm^{-2}$。因此，我们推断纯 $BiVO_4$ 不能输出恒定值的稳定光电流。如图 5-2（c）的电流-时间（i-t）曲线所示，尽管掺 Fe 的 $BiVO_4$ 具有较小的能隙，但掺 W 的 $BiVO_4$ 表现出更高的光电转换性能，这可能是因为掺 W 可增强 $BiVO_4$ 的 n 型特性，从而有效地增强载流子密度[25, 29]。但是，增加在稳定性测试中的辐照时间[图 5-2（d）]，很明显看到原始和 W 掺杂的 $BiVO_4$ 光电流急剧下降。因此，尽管进行了 W 掺杂可以显著提高光电转换性能，但不能提高 $BiVO_4$ 的 PEC 稳定性问题。W 掺杂是一种典型的 n 型掺杂，可替代一小部分 V^{5+} 位点。为了解决 $BiVO_4$ 的稳定性问题，我们最初提出用少量 Fe^{3+} 取代 Bi^{3+} 位点，光电流在短时间内稳定下来，然后在整个 PEC 测量期间保持固定值[图 5-2（d）]。因此，可以得出结论，铁掺杂可以极大地提高 $BiVO_4$ 的 PEC 稳定性，这可能是由于用 Fe^{3+} 替代 Bi^{3+} 可以改善 $BiVO_4$ 的晶体结构并消除 Bi^{3+} 引起的晶体缺陷。

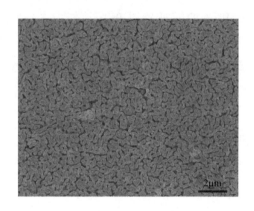

图 5-1　多孔 $BiVO_4$ 光电极的 SEM 图

为了改善 PEC 的稳定性和活性，我们设计了 Fe/W 共掺杂 $BiVO_4$ 光电极。图 5-3（a）给出了纳米孔 Fe/W 共掺杂 $BiVO_4$ 的 SEM 图像和元素图。可以看出，纳米多孔 $BiVO_4$ 的表面是相对光滑的，并且没有观察到其他次级纳米结构。另外，元素映射表明 Fe 和 W 元素的分布是均匀的。为了进一步揭示 Fe 和 W 离子的含量和分布，将基于 $BiVO_4$ 的透射电子显微镜（TEM）图像进行了能量色散 X 射线光谱（EDS）分析[图 5-3（b）和图 5-4]。由图 5-3（b）发现，MIL-100（Fe）中 160nm 范围内的黄色直线穿过单个 Fe/W 共掺杂的 $BiVO_4$ 纳米颗粒。与 Bi、V 和 O 元素的强信号相比，Fe 和 W 在 $BiVO_4$ 区域

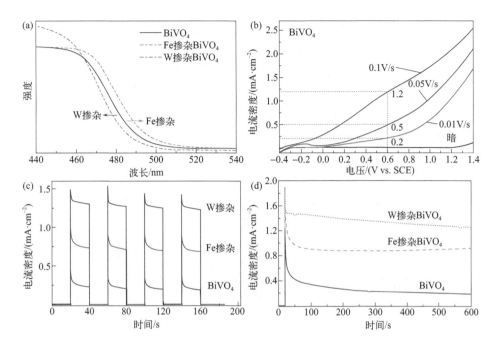

图 5-2 紫外-可见吸收光谱（a）、LSV（b）和在模拟下纯 $BiVO_4$、Fe 和 W 掺杂的 $BiVO_4$ 的 i-t 曲线（c）以及 i-t 稳定性测试（d）

仅显示非常弱的峰，表明微量存在。这与 Fe/W 共掺杂 $BiVO_4$ 的受控制造方法是一致的，Fe/W 的化学计量比仅为 3%。在 HRTEM 中观察到的 Fe/W 共掺杂 $BiVO_4$ 的 0.309nm 晶格条纹间距［图 5-3（c）］与纯 $BiVO_4$ 的（121）晶面一致[44]，这表明少量的 Fe 和 W 掺杂不会对 $BiVO_4$ 的晶体结构产生太大影响。获得掺杂的 $BiVO_4$ 光电极后，首先将包含 MIL-100（Fe）纳米粒子的溶液旋涂在 Fe/W 共掺杂的 $BiVO_4$ 膜上，然后在 150℃下干燥 30min。图 5-3（d）的 SEM 图像表明 MIL-100（Fe）纳米粒子分布在掺杂的 $BiVO_4$ 光电极的表面上。

研究了制备 $BiVO_4$ 的晶体结构和组成及 Fe/W 共掺杂 $BiVO_4$ 光电极在加载 MIL-100（Fe）之前和之后的 X 射线衍射（XRD）图谱，结果如图 5-5（a）所示。纳米多孔 $BiVO_4$ 光电极的 XRD 图谱与纯单斜晶体 $BiVO_4$ 结构的 XRD 图谱（JCPDS 14-0688）非常吻合[12-13]。此外，掺杂离子后未观察到显著变化，表明离子 Fe/W 共掺杂既不会影响 $BiVO_4$ 的晶体结构，也不会引入任何杂质。然而，通过物理旋涂法将 MIL-100（Fe）沉积在 Fe/W 共掺杂 $BiVO_4$ 薄膜上后，

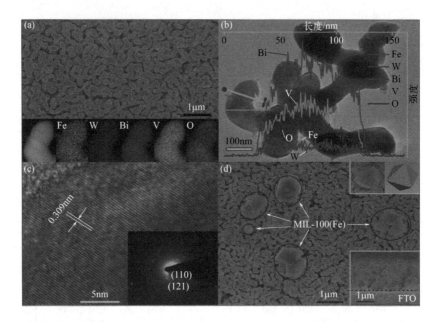

图 5-3　Fe/W 共掺杂 BiVO₄ 的 SEM 图像（插图是元素图）(a)、EDS 线分析 (b) 和 HRTEM 和选择区域衍射 (SAED)(c) 以及用 MIL-100 (Fe) 改性的 Fe/W 共掺杂 BiVO₄ 的 SEM 图像和截面图 (d)

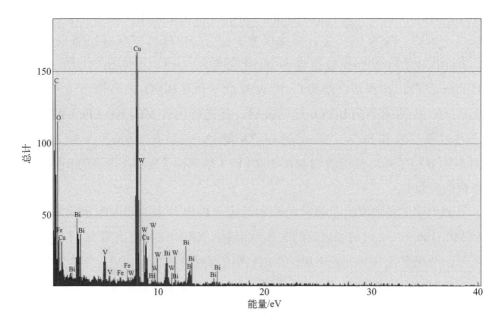

图 5-4　Fe/W 共掺杂 BiVO₄ 光电极的 EDS 谱图

在 2θ=11°附近出现了一个小的衍射峰。与纯 MIL-100（Fe）的 XRD 模式相比（图 5-6），可以得出结论，该小峰属于 MIL-100（Fe）。这些样品的紫外 - 可见吸收光谱显示在图 5-5（b）中。Fe/W 共掺杂的 $BiVO_4$ 的吸收边缘的红移可以归因于 Fe 掺杂，这是由 Fe 和 W 掺杂对 $BiVO_4$ 的吸收性能的影响而得出的 [图 5-2（a）]。由于 MIL-100（Fe）的吸收范围较大，通过在 Fe/W 共掺杂 $BiVO_4$ 膜上装饰 MIL-100（Fe），可以获得更大的红移。由吸收光谱计算出的这些样品的能隙已经显示在图 5-5（b）的插图中，MIL-100（Fe）的带隙仅约为 2eV。

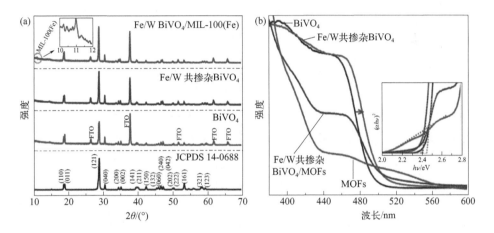

图 5-5　纯 $BiVO_4$ 的 XRD 图（a）和在装饰 MIL-100（Fe）之前和之后 Fe/W 共掺杂 $BiVO_4$ 的紫外 - 可见吸收光谱（b）

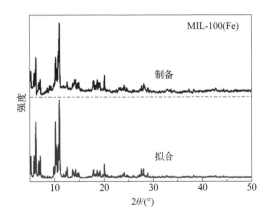

图 5-6　拟合和制备的 MIL-100（Fe）的 XRD 图

研究了 Fe/W 共掺杂 BiVO$_4$ 负载 MIL-100（Fe）改性后的 PEC 性能。图 5-7（a）展示了在模拟太阳光照射下相对于 SCE 的电位范围为 0.4～1.4 V 的 LSV。在各种偏置电压下，装饰后的 BiVO$_4$ 的光电流密度比纯 BiVO$_4$ 的高得多。电流法 i-t 图 [图 5-7（b）] 中的曲线清楚地表明在改性前后，电流密度分别为 1.43 和 2.76mA·cm^{-2}，是纯 BiVO$_4$（0.2mA·cm^{-2}）的 7 倍和 14 倍。负载 MIL-100（Fe）可以显著增强光电极的 PEC 性能，这可以通过纯 BiVO$_4$ 的 PEC 性能进一步证实 [图 5-8]。此外，与图 5-2（c）中的 Fe 或 W 单独掺杂的 BiVO$_4$ 相比，共掺杂的 BiVO$_4$ 表现出更高的光转换性能，表明 Fe 掺杂可以与 W 掺杂协同作用，从而进一步提高 BiVO$_4$ 的 PEC 活性。更重要的是，Fe 掺杂可以从根本上解决 BiVO$_4$ 光电极的 PEC 稳定性问题。如图 5-7（c），在整个稳定性测试中，BiVO$_4$ 显示出稳定的光电流输出。可以推论得出：如果仅采用铁掺杂，BiVO$_4$ 始终可以显示出高度稳定的 PEC 性能。这些样品的入射光电流转换效率（IPCE）光谱已测量，并且改性后的 BiVO$_4$ 在 420nm 处的 IPCE 最高为 52.6%，表明借助 MIL-100（Fe）改善了光生载流子的分离。

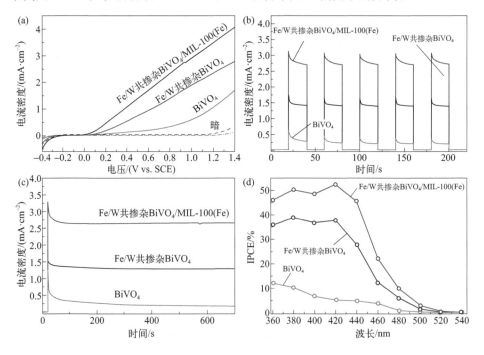

图 5-7　LSV（a）；电流法 i-t 曲线（b）；i-t 稳定性测试（c）；在模拟太阳光下纯 BiVO$_4$ 和在负载 MIL-100（Fe）前后情况下 BiVO$_4$ 的 IPCE 光谱（d）

为了探索增强的 PEC 性能的起源并发现界面电荷转移的性质，对 BiVO$_4$ 电化学阻抗谱（EIS）进行了测量，测量有和没有沉积 MIL-100（Fe）的 BiVO$_4$ 的 EIS，设置频率范围为 0.1～10^{5}Hz，电压为 0.6V，交流幅值为 5mV，测试电化学阻抗谱（EIS）奈奎斯特图如图 5-8（a）所示。选择 Randles–Ershler 模型作为等效电路，其中 R_s 是溶液电阻，CPE 是半导体/电解质界面的电容相位元素，R_{ct} 是跨电极的电荷转移电阻。奈奎斯特图中的拱形代表半圆直径，反映工作电极上的电荷转移动力学，反映了电荷转移电阻率[45-46]。如图 5-8（a）所示，改性后的 BiVO$_4$ 的阻抗电弧较小，比原始 BiVO$_4$ 的半径大。BiVO$_4$ 和负载 MIL-100（Fe）前后的 BiVO$_4$ 的拟合 R_{ct} 值分别为 484740、68339 和 32541Ω。因此，用 MIL-100（Fe）修饰的 BiVO$_4$ 具有较低的电荷转移电阻，具有比其他样品更高的电子迁移率，这归因于其更高的 PEC 活性。

此外，通常用莫特-肖特基（Mott–Schottky）分析测量并估计平带电势（E_{fb}）和电子载流子密度。如图 5-8（b）所示，可以使 BiVO$_4$ 的 E_{fb} 从 0.60V 移动至 0.62 V，使用 Mott-Schottky 可以将其进一步移至 0.72V，这有助于半导体/电解质界面处的电荷转移。在莫特-肖特基方程中，载流子密度与斜率成反比[47-48]，因此可以根据斜率计算出改性的 BiVO$_4$ 的最大载流子密度。因此，可以推断出 Fe/W 共掺杂的主要作用是降低 R_{ct}，会大大降低电荷转移过程中的界面阻力。负载 MIL-100（Fe）的主要作用是促进 E_{fb} 蓝移，有效地促进电荷分离，并间接证明 MIL-100（Fe）在整个催化剂系统中充当助催化剂。区分 Fe 和 W 掺杂在改善 BiVO$_4$ 的 PEC 活性中的作用，使用在 CASTEP 中实现的具有 PBE 功能的 DFT 计算，计算出纯 BiVO$_4$ 和共掺杂 BiVO$_4$ 的能隙结构[图 5-8（c）][49]。如图 5-8（d）所示，纯 BiVO$_4$ 电子中只能被大于带隙的能量从价带（O 2p）中激发到导带（V 3d）中。可以看作在 BiVO$_4$ 的禁带内形成杂质能带。值得注意的是，它完全由 Fe 3d 轨道组成，几乎没有观察到 W 的信号。在 BiVO$_4$ 的预定能隙内形成的 Fe 3d 能级缩小了 BiVO$_4$ 能隙的吸收，与图 5-5（b）中的紫外-可见光吸收率一致。因此，我们可以得出以下结论：Fe 掺杂的作用一方面是改善 PEC 的稳定性，另一方面是扩大光吸收区域。另外，W 掺杂是典型的 n 型掺杂，会导致靠近半导体/液体界面形成电场，因此有助于光生载流子的分离[25]。

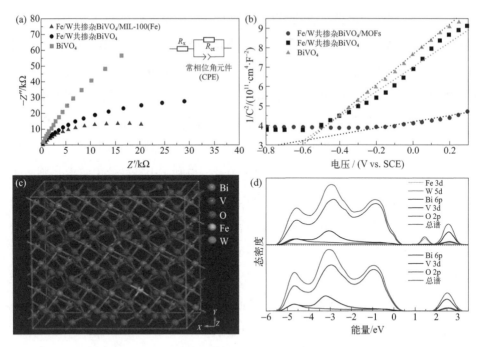

图 5-8 EIS 光谱的奈奎斯特图（插图显示了等效电路）（a）、以 5 kHz 的频率收集的 Mott-Schottky 图（b）BiVO$_4$ 的晶体结构（c）和计算的纯 BiVO$_4$ 和 Fe/W 共掺杂 BiVO$_4$ 的总态密度和偏态密度（d）

以前从未研究过负载 MIL-100（Fe）对光电极 PEC 性能的具体影响。由此，测试了 MIL-100（Fe）的光转换活性 [图 5-9（a）]。尽管 MIL-100（Fe）在紫外-可见吸收光谱中的吸收范围最宽 [图 5-5（b）]，但在模拟的太阳光照射下，它显示出最低的光电流密度。考虑 FTO 基板的光转换性能，负载 MIL-100（Fe）后 BiVO$_4$ 的 PEC 性能增强不是光转换的结果。考虑到 MIL-100（Fe）的低 PEC 性能以及对 Fe/W 掺杂的 BiVO$_4$ 的平带电势（E_{fb}）的负移影响较大，它只能作为 BiVO$_4$ 的助催化剂。

为进一步阐明 MIL-100（Fe）的功能拓展及其应用范围，还研究了用 MIL-100（Fe）修饰的 BiVO$_4$ 的 PEC 活性（图 5-10）。实验结果表明：作为助催化剂，MIL-100（Fe）只能增强 BiVO$_4$ 的光转换性能，而不能提高其稳定性。图 5-9（b）展示了用 MIL-100（Fe）改性的 BiVO$_4$ 的电荷转移机理。在光照下，BiVO$_4$ 中的光激发电子-空穴对可以在 BiVO$_4$ 和液体的界面处快速分离。空穴迅速转移到 MIL-100（Fe）上并引起氧化反应，而电子迁移到 FTO

上并触发对电极的还原反应。此外，人们还认为，MIL-100（Fe）独特的多孔结构有助于其与电解质的充分接触，因此有利于电荷的转移和运输[38, 49]，从而有助于增强 PEC 性能。因此，MIL-100（Fe）的沉积可以有效地促进电子-空穴对的分离，从而提高 $BiVO_4$ 的 PEC 活性。

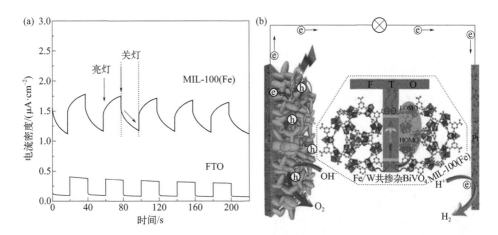

图 5-9　纯 MIL-100（Fe）在模拟太阳灯下的安培电流曲线（a）和 MIL-100（Fe）改性的 $BiVO_4$ 中的电荷分离和转移示意图（b）

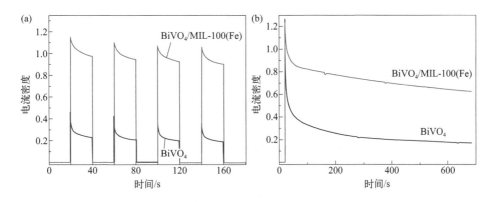

图 5-10　安培 i-t 曲线（a）模拟太阳灯下 $BiVO_4$ 和 $BiVO_4$/MIL-100（Fe）i-t 稳定性实验（b）

PEC 实验后对 Fe/W 共掺杂 BiVO4/MIL-100（Fe）光催化材料进行了 SEM 和 XRD 测试。如图 5-11（a）所示，不同尺寸的 MIL-100（Fe）纳米颗粒依然负载在 Fe/W 共掺杂 $BiVO_4$ 光电极上。Fe/W 共掺杂 $BiVO_4$/MIL-100

（Fe）光催化材料的 XRD 图也证实了这点［图 5-11（b）］。采用水热法制备的 MIL-100（Fe）不会在电解质溶液中溶解。另外在 Fe/W 共掺杂 $BiVO_4$ 光电极上负载 MIL-100（Fe）后将其放入烘箱中进行干燥，这可能会加强 MIL-100（Fe）在光电极上的附着。所以在 Fe/W 共掺杂 $BiVO_4$ 光电极上负载 MIL-100（Fe）是提高其 PEC 活性的一种有效可行的方法。

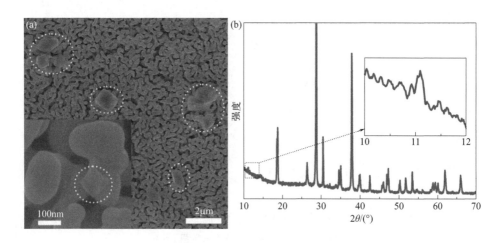

图 5-11　PEC 实验后 Fe/W 共掺杂 $BiVO_4$/MIL-100（Fe）光催化材料的 SEM 图（a）和 XRD 图（b）

5.4　结论

通过掺杂 Fe^{3+} 替代 Bi^{3+} 成功解决了 $BiVO_4$ 的光电化学（PEC）稳定性问题。Fe 掺杂可通过形成杂质带有效地缩小 $BiVO_4$ 的能隙，并与其他手段协同作用以进一步增强 $BiVO_4$ 的 PEC 性能，同时保持高稳定性。在本文中，我们设计和制备了经 MIL-100（Fe）改性的纳米多孔 $BiVO_4$，展现了其高 PEC 稳定性和活性。在这个新颖的系统中，Fe、W 和 MIL-100（Fe）分别充当稳定剂、n 型掺杂剂和助催化剂。Fe/W 掺杂的主要作用是降低跨界面的电荷转移电阻。密度泛函理论（DFT）计算清楚地证明了 Fe 和 W 掺杂在改善 $BiVO_4$ 的 PEC 性能方面的不同作用。此外，考虑到 MIL-100（Fe）量少、低 PEC 性能以及对 Fe/W 掺杂的 $BiVO_4$ 的平带电势（E_{fb}）的负移的影响较大，MIL-100（Fe）起着助催化剂的作用。

参考文献

[1] Fujishima A, Honda K. Electrochemical photolysis of water at a semiconductor electrode[J]. Nature, 1972, 238: 37-38.

[2] Liu S, Yu J, Jaroniec M. tunable photocatalytic selectivity of hollow TiO_2 microspheres composed of anatase polyhedra with exposed {001} facets[J]. J Am Chem Soc, 2010, 132: 11914-11916.

[3] Tang J, Cowan A, Durrant J, et al. Mechanism of O_2 production from water splitting: Nature of charge carriers in nitrogen doped nanocrystalline TiO_2 films and factors limiting O_2 production[J]. J Phys Chem C, 2011, 115: 3143-3150.

[4] Zheng Z, Huang B, Qin X, et al. Facile in situ synthesis of visible-light plasmonic photocatalysts M@TiO_2 (M=Au, Pt, Ag) and evaluation of their photocatalytic oxidation of benzene to phenol[J]. J Mater Chem, 2011, 21: 9079-9087.

[5] Liu L, Ouyang S, Ye J. Gold-nanorod-photosensitized titanium dioxide with wide-range visible-light harvesting based on localized surface plasmon resonance[J]. Angew Chem, 2013, 125: 6821-6825.

[6] Chen X, Burda C. The electronic origin of the visible-light absorption properties of C-, N- and S-doped TiO_2 nanomaterials[J]. J Am Chem Soc, 2008, 130: 5018-5019.

[7] Feng N, Wang Q, Zheng A, et al. Understanding the high photocatalytic activity of (B, Ag)-codoped TiO_2 under solar-light irradiation with XPS, solid-state NMR, and DFT calculations[J]. J Am Chem Soc, 2013, 135: 1607-1616.

[8] Ghicov A, Macak J, Tsuchiya H, et al. Ion implantation and annealing for an efficient N-doping of TiO_2 nanotubes[J]. Nano Lett, 2006, 6: 1080-1082.

[9] Ma S, Hisatomi T, Maeda K, et al. Enhanced water oxidation on Ta_3N_5 photocatalysts by modification with alkaline metal salts[J]. J Am Chem Soc, 2012, 134: 19993-19996.

[10] Li Y, Takata T, Cha D, et al. vertically aligned Ta_3N_5 nanorod arrays for solar-driven photoelectrochemical water splitting[J]. Adv Mater, 2013, 25: 125-131.

[11] Sun Y, Wu C, Long R, et al. Synthetic loosely packed monoclinic $BiVO_4$ nanoellipsoids with novel multiresponses to visible light, trace gas and temperature[J]. Chem Commun, 2009, 30:

4542-4544.

[12] Jo W, Jang J, Kong K, et al. phosphate doping into monoclinic BiVO$_4$ for enhanced photoelectrochemical water oxidation activity[J]. Angew Chem Int Ed, 2012, 51: 3147-3151

[13] Hong S, Lee S, Jang J, et al. Heterojunction BiVO$_4$/WO$_3$ electrodes for enhanced photoactivity of water oxidation[J]. Energy Environ Sci, 2011, 4: 1781-1787.

[14] Xi G, Ye J. Synthesis of bismuth vanadate nanoplates with exposed {001} facets and enhanced visible-light photocatalytic properties[J]. Chem Commun, 2010, 46: 1893-1895.

[15] McDonald K, Choi K. A new electrochemical synthesis route for a BiOI electrode and its conversion to a highly efficient porous BiVO$_4$ photoanode for solar water oxidation[J]. Energy Environ Sci, 2012, 5: 8553-8557.

[16] Yin C, Zhu S, Chen Z, et al. One step fabrication of C-doped BiVO$_4$ with hierarchical structures for a high-performance photocatalyst under visible light irradiation[J]. J Mater Chem A, 2013, 1: 8367-8378.

[17] Usai S, Obregn S, Becerro A, et al. monoclinic-tetragonal heterostructured BiVO$_4$ by yttrium doping with improved photocatalytic activity[J]. J Phys Chem C, 2013, 117: 24479-24484.

[18] Wang Z, Luo W, Yan S, et al. BiVO$_4$ nano-leaves: Mild synthesis and improved photocatalytic activity for O$_2$ production under visible light irradiation[J]. CrystEng Comm 2011, 13: 2500-2504.

[19] Kim T, Choi K. nanoporous BiVO$_4$ photoanodes with dual-layer oxygen evolution catalysts for solar water splitting[J]. Science, 2014, 343: 990-994.

[20] Berglund S, Flanherty D, Hahn N, et al. photoelectrochemical oxidation of water using nanostructured BiVO$_4$ films[J]. J Phys Chem C, 2011, 115: 3794-3802.

[21] He H, Berglund S, Rettie A, et al. Nanostructured Bi$_2$S$_3$/WO$_3$ heterojunction films exhibiting enhanced photoelectrochemical performance[J]. J Mater Chem A, 2013, 2: 9371-9379.

[22] Zhou M, Bao J, Bi W, et al. Efficient water splitting via a heteroepitaxial BiVO$_4$ photoelectrode decorated with Co-Pi catalysts[J]. ChemSusChem, 2012, 5: 1420-1425.

[23] Rao P, Cai L, Liu C, et al. Simultaneously efficient light absorption and charge separation in WO$_3$/BiVO$_4$ core/shell nanowire photoanode for photoelectrochemical water oxidation[J]. Nano Lett, 2014, 14: 1099-1105.

[24] Chang X, Wang T, Zhang P, et al. enhanced surface reaction kinetics and charge separation of p-n heterojunction Co_3O_4/$BiVO_4$ photoanodes[J]. J Am Chem Soc, 2015, 137: 8356-8359.

[25] Pilli S, Furtak T, Brown L, et al. Herring, cobalt-phosphate (Co-Pi) catalyst modified Mo-doped $BiVO_4$ photoelectrodes for solar water oxidation[J]. Energy Environ Sci, 2011, 4: 5028-5034.

[26] Berglund S, Rettie A, Hoang S, et al. Incorporation of Mo and W into nanostructured $BiVO_4$ films for efficient photoelectrochemical water oxidation[J]. Phys Chem Chem Phys, 2012, 14: 7065-7075.

[27] Luo W, Wang J, Zhao X, et al. Formation energy and photoelectrochemical properties of $BiVO_4$ after doping at Bi^{3+} or V^{5+} sites with higher valence metal ions[J]. Phys Chem Chem Phys, 2013, 15: 1006-1013.

[28] Ye H, Park H, Bard A. Screening of electrocatalysts for photoelectrochemical water oxidation on W-doped $BiVO_4$ photocatalysts by scanning electrochemical microscopy[J]. J Phys Chem C, 2011, 115: 12464-12470.

[29] Zhong D, Choi S, Gamelin D. Near-complete suppression of surface recombination in solar photoelectrolysis by "Co-Pi" catalyst-modified W: $BiVO_4$[J]. J Am Chem Soc, 2011, 133: 18370-18377.

[30] Zhang K, Shi X, Kim J, et al. Hotoelectrochemical cells with tungsten trioxide/Mo-doped $BiVO_4$ bilayers[J]. Phys Chem Chem Phys, 2012, 14: 11119-11124.

[31] Seabold J, Choi K. Efficient and stable photo-oxidation of water by a bismuth vanadate photoanode coupled with an iron oxyhydroxide oxygen evolution catalyst[J]. J Am Chem Soc, 2012, 134: 2186-2192.

[32] Jang J, Ham D, Narayanan L, et al. Role of platinum-like tungsten carbide as cocatalyst of CdS photocatalyst for hydrogen production under visible light irradiation[J]. Appl Catal A, 2008, 346: 149-154.

[33] Xiang Q, Yu J, Jaroniec M. Synergetic effect of MoS_2 and graphene as cocatalysts for enhanced photocatalytic H_2 production activity of TiO_2 nanoparticles[J]. J Am Chem Soc, 2012, 134: 6575-6578.

[34] Zong X, Han J, Ma G, et al. Photocatalytic H_2 evolution on CdS loaded with WS_2 as cocatalyst

under visible light irradiation[J]. J Phys Chem C, 2011, 115: 12202-12208.

[35] Arai T, Horiguchi M, Yanagida M, et al. Reaction mechanism and activity of WO_3-catalyzed photodegradation of organic substances promoted by a CuO cocatalyst[J]. J Phys Chem C, 2009, 113: 6602-6609.

[36] Yu H, Liu R, Wang X, et al. Enhanced visible-light photocatalytic activity of Bi_2WO_6 nanoparticles by Ag_2O cocatalyst[J]. Appl Catal B, 2012, 111-112: 326-333.

[37] Asai R, Nemoto H, Jia Q, et al. A visible light responsive rhodium and antimony-codoped $SrTiO_3$ powdered photocatalyst loaded with an IrO_2 cocatalyst for solar water splitting[J]. Chem Commun, 2014, 50: 2543-2546.

[38] Zhan W, Kuang Q, Zhou J, et al. Semiconductor@ metal-organic framework core-shell heterostructures: A case of ZnO@ ZIF-8 nanorods with selective photoelectrochemical response[J]. J Am Chem Soc, 2013, 135: 1926-1933.

[39] Zhang C, Qiu L, Zhu F, et al. A novel magnetic recyclable photocatalyst based on a core-shell metal-organic framework Fe_3O_4@MIL-100(Fe) for the decolorization of methylene blue dye[J]. J Mater Chem, 2013, 1: 14329-14334.

[40] Horcajada P, Surblé S, Serre C, et al. Synthesis and catalytic properties of MIL-100(Fe), an iron(Ⅲ) carboxylate with large pores[J]. Chem Commun, 2007, 27: 2820-2822.

[41] Perdew J, Burke K, Ernzerhof M. Generalized gradient approximation made simple[J]. Phys Rev Lett, 1996, 77: 3865-3868.

[42] Clark S, Segall M, Pickard C, et al. First principles methods using CASTEP[J]. 2005, 220: 567-570.

[43] Feng C, Jiao Z, Li S, et al. Facile fabrication of $BiVO_4$ nanofilms with controlled pore size and their photoelectrochemical performances[J]. Nanoscale, 2015, 7: 20374-20379.

[44] Li C, Wang S, Wang T, et al. Monoclinic porous $BiVO_4$ networks decorated by discrete g-C_3N_4 nano-islands with tunable coverage for highly efficient photocatalysis[J]. Small, 2014, 10: 2783-2790.

[45] Kim H, Mora-Sero I, Gonzalez-Pedro V, et al. Mechanism of carrier accumulation in perovskite thin-absorber solar cells[J]. Nat Commun, 2013, 4: 2242.

[46] Shen L, Uchaker E, Zhang X, et al. Hydrogenated $Li_4Ti_5O_{12}$ nanowire arrays for high rate

lithium ion batteries[J]. Adv Mater, 2012, 24: 6502-6506.

[47] Liao L, Zhang Q, Su Z, et al. Efficient solar water-splitting using a nanocrystalline CoO photocatalyst[J]. Nat Nanotechnol, 2014, 9: 69-73.

[48] Mao C, Zuo F, Hou Y, et al. In situ preparation of a Ti^{3+} self-doped TiO_2 film with enhanced activity as photoanode by N_2H_4 reduction[J]. Angew Chem, 2014, 126: 10653-10657.

[49] Wang J, Wang C, Lin W. Metal-organic frameworks for light harvesting and photocatalysis[J]. ACS Catal, 2012, 2: 2630-2640.

第六章

纳米多孔 $FeVO_4/g-C_3N_4$ 复合光催化材料制备及性能研究

6.1	引言	149
6.2	实验部分	149
6.3	结果与讨论	151
6.4	结论	154
	参考文献	154

6.1 引言

光电化学（PEC）裂解水制氢是一种有前途的清洁能源可再生方法，可解决能源危机[1-4]。对于传统半导体，如 TiO_2、WO_3、ZnO 和 $TaON$，宽带隙和低电荷分离效率严重限制了它们的实际应用[5-7]。因此，具有可见光活性和稳定性的光电极的探索和开发激发了人们对 PEC 领域的研究兴趣。最近，由地球上储藏量丰富的元素 Fe 和 V 组成的 $FeVO_4$，由于其良好的带隙宽度（2.06eV）、足够低的价带位置和优异的化学稳定性，引起了人们的极大兴趣[8-9]。析氧反应（OER）由于反应动力学缓慢成为诸多能源转化及储存过程中的瓶颈步骤。为了解决这个问题，金属离子掺杂、异质复合和助催化剂沉积等各种策略已被广泛应用于提高 $FeVO_4$ 的 PEC 性能中[10-12]。对于助催化剂沉积方法，具有可逆氧化还原状态的 p 型Ⅷ金属（Fe、Co、Ni）氢氧化物和氧化物都可以有效地提高光电极的光电化学性能[12-13]。但是大量使用具有明显毒性的金属离子肯定与环境保护的趋势背道而驰。此外，大块助催化剂的大厚度或尺寸可能会阻碍半导体对阳光的吸收或延长空穴转移距离。因此，迫切需要探索一种无毒的超薄结构的助催化剂来改善 $FeVO_4$ 的 PEC 性能。在这项工作中，我们首先将多孔 $FeVO_4$ 光电极与超薄 $g-C_3N_4$ 纳米片复合，以通过简单的浸渍方法提高 PEC 性能。$g-C_3N_4$ 作为一种非金属多聚体光催化剂，由于其在可见光下优异的光催化析氢性能，引起了人们的极大兴趣[14-15]。人们发现超薄的 $g-C_3N_4$ 可以有效地降低电荷复合效率，并且存储在价带中的光生空穴可以及时参与水氧化过程。奈奎斯特（Nyquist）曲线进一步证明超薄 $g-C_3N_4$ 纳米片可以显著增加载流子密度并促进电子-空穴对分离。因此，与纯 $FeVO_4$ 光电极相比，$FeVO_4/C_3N_4$ 异质结构在光电流密度和起始电位方面表现出更好的 PEC 性能。

6.2 实验部分

6.2.1 $FeVO_4$ 的合成

以 $Fe(NO_3)_3 \cdot 9H_2O$、NH_4VO_3 和 $C_6H_8O_7$ 为原料合成了 $FeVO_4$。具体步骤如下，将 1.5mmol $Fe(NO_3)_3 \cdot 9H_2O$ 和 0.5 mmol $C_6H_8O_7$ 溶解在 20mL

蒸馏水中，并在80℃下搅拌5min，使其完全溶解。类似地，溶解1.5mmol NH_4VO_3（约80℃，＜5min）。之后将 NH_4VO_3 加入 $Fe(NO_3)_3 \cdot 9H_2O$ 溶液中，将混合物在80℃下保持30min，迅速形成黄色纳米颗粒胶体。接下来，将0.12g PEG-600溶解在10mL乙醇中，然后转移到 $FeVO_4$ 溶液中，为了增强涂层的润湿性和均匀性，在室温下摇晃1.0h，搅拌溶液直至完全反应，胶体明显变黄。然后，将 $FeVO_4$ 前驱体溶液（200μL）滴铸在之前已清洁的FTO衬底上。样品在80℃的烘箱中干燥55min，最后在500℃的空气中退火2.5h，在FTO衬底上获得橙色薄膜。

6.2.2 块状 g-C_3N_4 的合成

通过以下步骤合成块状g-C_3N_4，将10g双氰胺放入带盖的坩埚中，并在马弗炉中以约2.3℃/min的升温速率加热至550℃保温4h。随后将获得的黄色产物（约5g）彻底研磨，以便进一步加工和表征。

6.2.3 超薄 g-C_3N_4 纳米片的合成

根据先前报道的方法合成了超薄g-C_3N_4纳米片，该方法使用块状g-C_3N_4作为前驱体。具体的合成步骤是，将4.0 g 块状g-C_3N_4、52g H_2SO_4（98%）和20g发烟硫酸（含游离SO_3 20%～25%）加入100mL烧瓶中，并在140℃下搅拌2h后在170℃搅拌3h。将获得的透明黄色溶胶自然冷却，然后在75℃及搅拌下注入800mL去离子水中，水逐渐变成淡白色的悬浮液。随后向溶液中加入85.5g NH_4Cl（1.60mol），并将溶液在70℃下搅拌2h，静置1h，然后热过滤以去除残留物。将获得的透明无色滤液（保持在70℃）快速置于冰浴中并搅拌1.5h，产生纯白色沉淀并通过过滤器收集，用去离子水和乙醇洗涤，最后在60℃下真空干燥。将所得粉末酸化为超薄g-C_3N_4纳米片。

6.2.4 $FeVO_4$/g-C_3N_4 异质结的制备

$FeVO_4$/g-C_3N_4异质结通过将$FeVO_4$光电极浸入超薄g-C_3N_4纳米片溶液（0.5mg/mL）中并在350℃下退火1h获得。

6.2.5 表征

XRD测量是在仪器上使用CuKα辐射（40kV）进行的。XRD扫描范围是从 10°到 90°,扫描速率为 0.0678s^{-1}。在场发射扫描电子显微镜上以 5kW 的加速电压进行 SEM 测量。TEM 测量是通过使用 200kW 的透射电子显微镜进行的。使用 BaSO$_4$ 作为背景,在紫外-可见光谱仪上获得紫外-可见光漫反射光谱。

6.2.6 光电化学性能

使用 FeVO$_4$/g-C$_3$N$_4$ 作为 PEC 电池中的光电阳极。光电流响应 CHI-660D 稳压器以 Pt 切片作为夹心型配置,在可见光照射下记录。对于电极,使用饱和甘汞电极(SCE)作为参比电极,以 0.1mol/L Na$_2$SO$_4$ 溶液作为电解质。装有模拟太阳滤光片将 300W 氙弧灯,模拟太阳滤光片校准为 100mW·cm^{-2},并使用辐射计作为光源。使用 300W Xe 灯和单色仪在 0.8V 下对 SCE 进行单色光下的光电流响应测量。莫特-斯科蒂图是使用电化学分析仪在标准三电极系统中以 5 kHz 的频率进行测试的。设置频率范围为 0.1~10^{5}Hz;电压为 0.6V,交流(AC)幅值为 5mV,测试电化学阻抗谱(EIS)奈奎斯特(Nyquist)图。所有实验均在室温条件下进行。

6.3 结果与讨论

以 Fe(NO$_3$)$_3$ 和 NH$_4$VO$_3$ 为反应物,柠檬酸为结构导向剂,通过简单的滴铸和热处理方法制备了在 FTO 衬底上生长的纳米多孔 FeVO$_4$ 光电极。通过场发射扫描电子显微镜(FESEM)观察所制备的 FeVO$_4$ 纳米膜的形貌。如图 6-1(a)所示:FeVO$_4$ 是由许多直径为 50~100nm 的相对均匀相互交叉的纳米颗粒构成的纳米多孔结构。此外,还通过高分辨率透射电子显微镜(HRTEM)观测了制备的 FeVO$_4$ 光电极的结构。如图 6-1(b)所示,由晶格条纹计算的间距值 d 为 0.32nm,对应的是(220)晶面[16]。厚度约为 2nm 的超薄 g-C$_3$N$_4$ 纳米片紧紧包覆在 FeVO$_4$ 纳米颗粒周围。

图 6-2(a)是纳米多孔 FeVO$_4$/g-C$_3$N$_4$ 异质结光催化材料的紫外-可见光吸收光谱。从图中可以看出:在 FeVO$_4$ 上负载 g-C$_3$N$_4$ 不会影响 FeVO$_4$ 的光吸收范围。为了进一步阐明 FeVO$_4$ 的晶体结构和组成,对样品进行了 X 射线衍射(XRD)测试。如图 6-2(b)所示,除了 FTO 的衍射峰外,所有特征峰均为正交相 FeVO$_4$(JCPDS,038-1372)的峰[17]。与纯纳米多孔 FeVO$_4$ 相比,可

以看出 FeVO$_4$/g-C$_3$N$_4$ 异质结光催化材料的 XRD 谱图没有明显差异，这意味着超薄 g-C$_3$N$_4$ 纳米片的分散均匀。

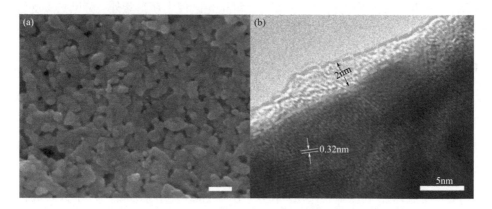

图 6-1　纳米多孔 FeVO$_4$/g-C$_3$N$_4$ 复合光催化材料的
SEM（a）和 HRTEM（b）图

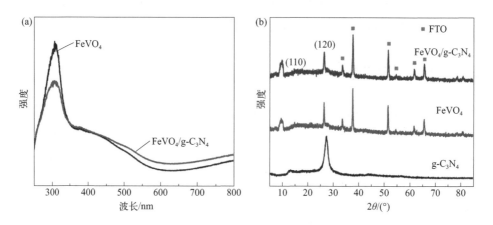

图 6-2　纳米多孔 FeVO$_4$/g-C$_3$N$_4$ 异质结光催化材料的紫外－可见光吸收光谱（a）
FeVO$_4$ 和 g-C$_3$N$_4$ 的 XRD 谱图（b）

如图 6-3 所示，在模拟太阳光照射下和 0.1mol/L Na$_2$SO$_4$ 溶液中，探索了合成的 FeVO$_4$ 和 FeVO$_4$/g-C$_3$N$_4$ 复合光电极的 PEC 性能。电流 i-t 曲线 [图 6-3（a）] 清楚地表明，FeVO$_4$/g-C$_3$N$_4$ 异质结的光电流密度是 0.5mA/cm^2，远高于 FeVO$_4$ 的 0.18 mA/cm^2，这可以通过图 6-3（b）中的斩波线性扫描伏安图（LSV）进一步证实。FeVO$_4$/g-C$_3$N$_4$ 光电极的光电流密度增强可归因于超薄 g-C_3N_4 层上的快速空穴俘获，从而促进了其有效电子 - 空穴对分离。此外，LSV 表明：FeVO$_4$/g-C_3N_4 异质结的起始电位略低于纯 FeVO$_4$。这表明 g-C_3N_4 有利于电荷分离，从而改善了 FeVO$_4$ 的 PEC 性能。图 6-3（c）展示了 $1/C^2$ 相对电位的 Mott-Schottky 曲线，其中斜率为正，表明 FeVO$_4$ 是 n 型半导体[18]。FeVO$_4$/g-C_3N_4 异质结的较小斜率说明了其比纯 FeVO$_4$ 光电极具有更大的载流子密度。此外，FeVO$_4$/g-C_3N_4 线与横轴红移的交点比 FeVO$_4$ 的稍微偏移，这与 LSV 曲线一致。如图 6-3（d）所示，为了探求 FeVO$_4$/g-C_3N_4 光电极 PEC 活性增强的原因和界面电荷转移特性，测试了这些样品的电化学阻抗谱（EIS）（电子频率为 0.1～10^5Hz，开路电位，5mV 的振幅），奈奎斯特图中 Randles-Ershler 等效电路模型拟合的电弧与电极/电解质界面处的电荷转移有关。可以看出，FeVO$_4$/g-C_3N_4 的半圆弧低于纯 FeVO$_4$ 光电极的，这表明超薄的 g-C_3N_4 结构有利于有效的电荷分离和迁移。

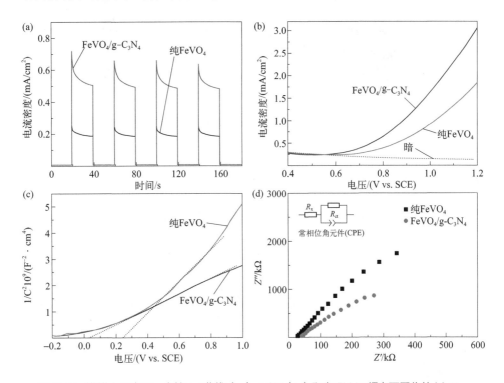

图 6-3　模拟太阳光下，安培 i-t 曲线（a）、LSV（b）和在 5kHz 频率下采集的 Mott-Schottky 曲线（c）以及电化学阻抗谱的 Nyquist 图（插图为等效电路）（d）

图 6-4 表明了超薄 g-C_3N_4 修饰的 $FeVO_4$ 光阳极中电荷转移的可能机制。在太阳光照射下，由于超薄 g-C_3N_4 的高价带位置，$FeVO_4$ 的光激发空穴可以迅速从 $FeVO_4$ 价带转移到 g-C_3N_4 价带，用于后续氧化反应，从而促进光生载流子分离并提高 $FeVO_4$ 的 PEC 性能。同时光生电子迁移到 FTO 并通过外部电路迁移到 Pt 电极，参与还原反应。值得注意的是，纳米多孔和超薄片结构材料的复合缩短了光生电子 - 空穴的行进距离，这提高了 $FeVO_4$ 的 PEC 活性并为 PEC 水氧化提供了足够的表面活性位点。

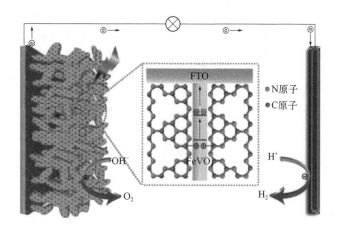

图 6-4 超薄 g-C_3N_4 修饰的 $FeVO_4$ 光电极中电荷分离和转移图

6.4 结论

总之，将超薄 g-C_3N_4 纳米片与纳米多孔 $FeVO_4$ 光电极复合是一种有效合理地提高 $FeVO_4$ 的 PEC 活性的方法。$FeVO_4$/g-C_3N_4 异质结显示出比纯纳米多孔 $FeVO_4$ 光电极高两倍以上的光电流密度，这可归因于超薄 g-C_3N_4 纳米片可有效捕获光激发空穴，从而促进光生载流子的分离并提高 $FeVO_4$ 光电极的 PEC 性能。

参考文献

[1] Hisatomi T, Kubota J, Domen K. Recent advances in semiconductors for photocatalytic and photoelectrochemical water splitting[J]. Chem Soc Rev, 2014, 43: 7520-7535.

[2] Nozik A. Photoslectrochemistry: Applications to solar energy conversion[J]. Annu Rev Phys Chem, 1978, 29: 189-222.

[3] Khaselev O, Turner J. Variations on a theme-recent developments on the mechanism of the heck reaction and their implications for synthesis[J]. Science, 1998, 280: 425-427.

[4] Yu H, Shi R, Zhao Y, et al. Defect-rich ultrathin ZnAl-layered double hydroxide nanosheets for efficient photoreduction of CO_2 to CO with water[J]. Adv Mater, 2016, 28: 9454-9477.

[5] Hu S, Shaner M, Beardslee J, et al. Amorphou TiO_2 coatings stabilize Si, GaAs, and GaP photoanodes for efficient water oxidation[J]. Science, 2014, 344: 1005-1009.

[6] Abe R, Higashi M, Domen K. Facile fabrication of an efficient oxynitride TaON photoanode for overall water splitting into H_2 and O_2 under visible light irradiation[J]. J Am Chem Soc, 2010, 132: 11828-11829.

[7] Pesci F, Cowan A, Alexander B, et al. Charge carrier dynamics on mesoporous WO_3 during water splitting[J]. J Phys Chem Lett, 2011, 2: 1900-1903.

[8] Biswas S, Baeg J. A facile one-step synthesis of single crystalline hierarchical WO_3 with enhanced activity for photoelectrochemical solar water oxidation[J]. Int J Hydrogen Energy, 2013, 38: 14451-14457.

[9] Tang D, Rettie A, Mabayoje O, et al. Facile growth of porous $Fe_2V_4O_{13}$ films for photoelectrochemical water oxidation[J]. J Mater Chem A, 2016, 4: 3034-3042.

[10] Zhou M, Bao J, Xu Y, et al. Photoelectrodes based upon Mo: $BiVO_4$ inverse opals for photoelectrochemical water splitting[J]. ACS Nano, 2014, 8: 7088-7098.

[11] Ma M, Kim J, Zhang K, et al. Double-deck inverse opal photoanodes: Efficient light absorption and charge separation in heterojunction[J]. Chem Mater, 2014, 26: 5592-5597.

[12] Chang X, Wang T, Zhang P, et al. Enhanced surface reaction kinetics and charge separation of p-n Heterojunction $Co_3O_4/BiVO_4$ photoanodes[J]. J Am Chem Soc, 2015, 137: 8356-8359.

[13] Zhong M, Hisatomi T, Kuang Y, et al. Surface modification of CoO_x loaded $BiVO_4$ photoanodes with ultrathin p-type NiO layers for improved solar water oxidation[J]. J Am Chem Soc, 2015, 137: 5053-5060.

[14] Wang X, Maeda K, Thomas A, et al. A metal-free polymeric photocatalyst for hydrogen production frorrr water under visible light[J]. Nat Mater, 2009, 8:76-80.

[15] Li Y, Ouyang S, Xu H, et al. Constructing solid-gas-interfacial fenton reaction over alkalinized-C_3N_4 photocatalyst to achieve apparent quantum yield of 49% at 420 nm[J]. J Am Chem Soc, 2016, 138: 13289-13297.

[16] Nithya V, Kalai Selvan R, Sanjeeviraja C, et al. Synthesis and characterization of $FeVO_4$ nanoparticles[J]. Mater Res Bull, 2011, 46:1654-1658.

[17] Wang W, Zhang Y, Wang L, et al. Facile synthesis of Fe^{3+}/Fe^{2+} self-doped nanoporous $FeVO_4$ photoanodes for efficient solar water splitting[J]. J Mater Chem A, 2017, 5: 2478-2482.

[18] Jiao Z, Shang M, Liu J, et al. The charge transfer mechanism of Bi modified TiO_2 nanotube arrays: TiO_2 serving as a "charge-transfer-bridge"[J]. Nano Energy, 2017, 31: 96-104.

第七章

铁掺杂纳米多孔 BiVO$_4$/MIL-53（Fe）复合光催化材料制备及性能研究

7.1 引言 158
7.2 实验部分 158
7.3 结果与讨论 160
7.4 结论 164
参考文献 165

7.1 引言

光电化学（PEC）裂解水作为一种有前景的将太阳能转化为化学能的策略已被广泛研究[1-3]。自首次使用 TiO_2 光电极进行水分解以来，科研工作者对能有效利用太阳能的光催化材料进行了大量探索[4]。然而，TiO_2 的宽带隙限制了其充分利用太阳能，仅能利用紫外波段范围的光[5]。为了更充分利用太阳能，PEC 领域的科研工作者将目光投向了可见光响应半导体。基于此，单斜钒酸铋（m $BiVO_4$）因其良好的带隙宽度、有吸引力的晶体结构和独特的性能而被认为是最有吸引力的可见光光催化剂之一[6-9]。$BiVO_4$ 被认为是一种有前途的光电极，因为，它的早期光电流阈值大约为 0.0V[10-12]。在低偏置电势下，它显示出比其他半导体更高的光电流密度。然而，$BiVO_4$ 的电荷分离和传输性能不足以及 PEC 稳定性差，极大地限制了其实际应用。科研工作者已经采取了各种方法来解决这些问题，包括离子掺杂、异质耦合和助催化剂负载[13-15]。但是提高 $BiVO_4$ 的 PEC 性能仍然是一个巨大挑战。本文中我们采用 Fe 离子掺杂的方法来改善 $BiVO_4$ 的 PEC 稳定性，即用 Fe 离子取代少量 Bi 离子。用 Fe 离子取代 Bi 离子将使光电流的值稳定，保持在恒定值，这可能是因为 Fe 离子掺杂能够克服 $BiVO_4$ 的结构缺陷。此外，Fe 掺杂不仅有利于 $BiVO_4$ 的 PEC 稳定性和光转化效率，而且可以与助催化剂协同作用，进一步提高 $BiVO_4$ 光电极的 PEC 性能。在这项工作中，我们负载了 MOFs 材料 MIL-53（Fe）作为助催化剂，以进一步提高光生载流子的分离效率（图 7-1）。实验结果表明，与 $BiVO_4$ 相比，用 MIL-53（Fe）改性的 Fe 掺杂 $BiVO_4$ 具有更佳的 PEC 性能。

7.2 实验部分

7.2.1 Fe 离子掺杂的 $BiVO_4$ 材料的制备

通过在 FTO 衬底上滴铸前驱体溶液制备了纳米多孔掺铁 $BiVO_4$ 光电极。首先，将 $Bi(NO_3)_3 \cdot 5H_2O$ 和 NH_4VO_3 分别溶解在乙二醇溶剂（75mL）中。将适量的 $Fe(NO_3)_3$ 与 $Bi(NO_3)_3$ 溶液混合用于 Fe 离子掺杂。然后将 0.68g PEG-600 溶解在另一 20mL 乙二醇溶剂中。其次，根据化学计量比，按以下配方混合三种溶液：5mL $Bi(NO_3)_3 \cdot 5H_2O$ 溶液、5mL NH_4VO_3 溶液和 2.5mL

PEG-600 溶液。再次，在 FTO 衬底上滴铸 0.2mL 前驱体溶液。最后，将样品在 150℃的烘箱中干燥 60min，然后在 500℃的马弗炉中退火 2.5h。

7.2.2 MIL-53（Fe）的制备

采用水热法制备了 MIL-53（Fe）。在 N, N-二甲基甲酰胺（DMF，5mL）的溶剂中混合等物质的量的氯化铁（$FeCl_3 \cdot 6H_2O$，1mmol）、对苯二甲酸（TPA，1mmol）和 HF（1mmol）。将前驱体溶液转移到聚四氟乙烯内衬不锈钢反应釜中，并在 150℃下加热 3d。用甲醇洗涤合成的黄色 MIL-53（Fe）固体，从材料的孔中除去 DMF 溶剂。然后用水 [1g MIL-53（Fe）溶于 0.5L 水中] 代替甲醇，之后自然干燥该黄色粉末。

7.2.3 Fe 离子掺杂 $BiVO_4$/MIL-53（Fe）复合材料的制备

首先将 MIL-53（Fe）分散到乙醇溶液中，然后使用旋涂法将其负载在掺杂 Fe 的 $BiVO_4$ 光电极上。最后将 MIL-53（Fe）改性的掺杂 Fe 的 $BiVO_4$ 在 150℃的烘箱中干燥 30min。

7.2.4 表征

X 射线衍射（XRD）光谱测量是在仪器上使用 CuKα 辐射（40kV）进行的，扫描范围为 10°到 90°，扫描速率为 $0.0678s^{-1}$。在场发射扫描电子显微镜上以 5kV 的加速电压进行 SEM 测量。TEM 测量是通过使用 200kV 的透射电子显微镜进行的。使用 $BaSO_4$ 作为背景，在紫外-可见光谱仪上获得紫外-可见光漫反射光谱，光谱范围为 250～800nm。

7.2.5 光电化学性能

将 MIL-53（Fe）修饰的掺杂 Fe 的 $BiVO_4$ 用作 PEC 电池中的光电极，模拟太阳光照射下的光电流响应。用 CHI-660D 恒电位仪以夹心型配置记录，使用 Pt 片作为对电极，饱和甘汞电极（SCE）作为参比电极，0.1mol/L Na_2SO_4 溶液作为电解质。光源为 300W 氙弧灯，配有校准至 100mW/cm^2 的模拟太阳滤光片，用辐射计测量。电化学阻抗谱（EIS）Nyquist 曲线在频率范围为 0.1～10^5Hz，电压为 0.6V，交流振幅为 5mV 下获得。所有实验都在室温条件下进行。

7.3　结果与讨论

多孔 $BiVO_4$ 膜是通过简单的滴铸法制备的。Fe 离子掺杂的 $BiVO_4$ 也使用相同的方法制备。如图 7-1 所示，将 MIL-53（Fe）分散到乙醇溶液中，然后使用旋涂法将其负载在掺杂 Fe 的 $BiVO_4$ 光电极上。

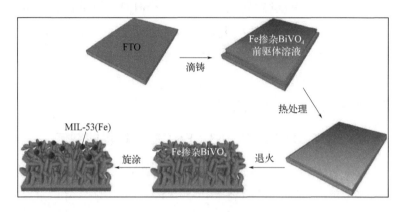

图 7-1　用 MIL-53（Fe）改性的 Fe 离子掺杂的 $BiVO_4$ 的制备工艺示意图

图 7-2（a）是 Fe 掺杂的 $BiVO_4$ 的扫描电子显微镜（SEM）图像。从图中可以看出纳米 $BiVO_4$ 多孔膜的表面相对光滑，没有其他二级纳米结构。图 7-2（b）是用 MIL-53（Fe）修饰的 Fe 掺杂的 $BiVO_4$ 的 SEM 图像，从图中可以看出与平滑的 Fe 掺杂的 $BiVO_4$ 相比，其表面有许多尺寸从 500nm 到 1μm 的纳米颗粒，这些纳米颗粒可以由元素映射图像确定为 MIL-53（Fe）。Fe 离子掺杂的 $BiVO_4$ 的元素映射图像如图 7-2（c）所示，除了 Bi、V、O 分散均匀之外，可以看出 Fe 元素也均匀分布在整个 $BiVO_4$ 纳米颗粒中，这意味着 Fe 掺杂的 $BiVO_4$ 组成的一致性。值得注意的是，图 7-2（d）不是图 7-2（b）中用 MIL-53（Fe）修饰的 Fe 掺杂的 $BiVO_4$ 的元素映射图像。因为 MIL-53（Fe）的组分也含有 Fe 元素，无法与 Fe 掺杂的 $BiVO_4$ 中的 Fe 元素区分。因此，为了阐明 MIL-53（Fe）的存在，我们还制备了用 MIL-53（Fe）修饰的 $BiVO_4$，其元素映射图像如图 7-2（d）所示。从图 7-2（d）中 Bi 和 V 的元素分布图像可以清楚地看出，纳米多孔结构为纯 $BiVO_4$，大纳米粒子为黑色，这意味着其中未检测到 Bi 和 V 元素。然而，由于 $BiVO_4$ 和 MIL-53（Fe）都含有 O 原子，所以 O 的整个元素映射图像是明亮的。此外只有纳米颗粒 Fe 元素映射显示是亮

的，基底是暗的，因为基底是没有 Fe 离子的纯 BiVO$_4$。因此，可以明确说明纳米多孔结构是 BiVO$_4$，其上修饰的纳米颗粒是 MIL-53（Fe），这与图 7-2（b）相一致。

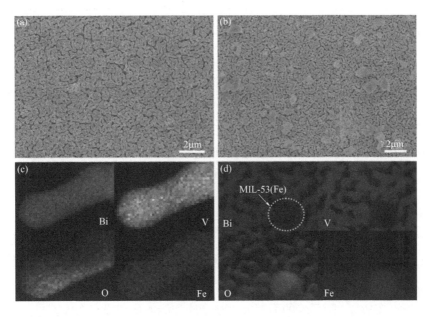

图 7-2　无 MIL-53（Fe）（a）和有 MIL-53（Fe）（b）修饰的掺铁 BiVO$_4$ 的 SEM 图像及用 MIL-53（Fe）修饰的 Fe 掺杂 BiVO$_4$（c）和纯 BiVO$_4$（d）的元素图

掺 Fe BiVO$_4$ 含 MIL-53（Fe）和无 MIL-53（Fe）的 UV/Vis 吸收光谱如图 7-3（a）所示。由于 MIL-53（Fe）的光吸收范围较大，在 Fe 掺杂的 BiVO$_4$ 膜上负载 MIL-53（Fe），其 UV/Vis 吸收光谱有较大的红移。为了进一步阐明这些制备的纳米颗粒的晶体结构和组成，还对样品进行了 XRD 测试。如图 7-3（b）所示，掺杂 Fe 的纳米多孔 BiVO$_4$ 光电极的 XRD 图谱与纯单斜 BiVO$_4$ 结构晶体的 XRD 图一致，掺杂 Fe 离子后未观察到明显变化，表明 Fe 掺杂既不会影响 BiVO$_4$ 的晶体结构，也不会引入任何杂质。然而通过物理旋涂法在掺杂 Fe 的 BiVO$_4$ 薄膜上沉积 MIL-53（Fe）后，位于 $2\theta=9°$ 附近的小衍射峰变宽，通过与纯 MIL-53（Fe）的 XRD 图谱比较，可以将其归因于 MIL-53（Fe）的添加。由此可以确认引入的 Fe 离子将拓宽 BiVO$_4$ 的光吸收范围，并且不会带来任何杂质。

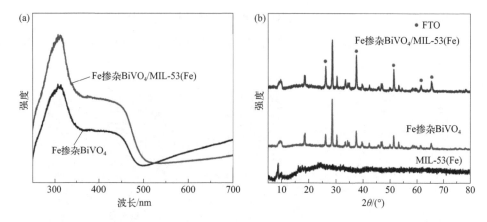

图 7-3　掺 Fe BiVO$_4$ 含 MIL-53（Fe）和无 MIL-53（Fe）的 UV/Vis 吸收光谱（a）和 XRD 谱图（b）

图 7-4（a）是 MIL-53（Fe）改性 Fe 掺杂的 BiVO$_4$ 的 XPS 光谱图。如图 7-4（a）所示，XPS 光谱图证实 Fe、Bi、V 和 O 元素的存在。图 7-4（b）是加入 MIL-53（Fe）前后 Fe 掺杂 BiVO$_4$ 的 Fe 2p XPS 光谱图。由图 7-4（b）可以看出，Fe 掺杂的 BiVO$_4$ 中 Fe 元素的含量远小于用 MIL-53（Fe）改性的 Fe 掺杂的 BiVO$_4$。除了峰的强度外，结合能和峰的形态没有太大差异。因此，可以推断 MIL-53（Fe）已成功地负载在 Fe 掺杂的 BiVO$_4$ 纳米多孔材料上了。

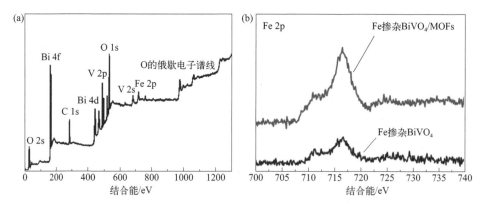

图 7-4　MIL-53（Fe）改性 Fe 掺杂的 BiVO$_4$ 的 XPS 光谱图（a）和含 MIL-53（Fe）和无 MIL-53（Fe）的掺 Fe BiVO$_4$ 的 Fe 2p XPS 光谱图（b）

为了提高 BiVO$_4$ 的稳定性和 PEC 性能，我们提出用 Fe 离子取代 Bi 离子，这可以改善 BiVO$_4$ 晶体结构并消除晶体缺陷。如图 7-5 所示，在

0.1mol/L 的 Na_2SO_4 溶液中，在太阳光照射下，研究了 $BiVO_4$、Fe 掺杂的 $BiVO_4$ 和 MIL-53（Fe）改性 Fe 掺杂的 $BiVO_4$ 的 PEC 性能。虽然 Fe 离子掺杂可以解决纳米多孔 $BiVO_4$ 的稳定性问题，但由于其固有的光生载流子复合率高，其光电流密度不能大幅度提高。为了进一步提高光生电荷分离效率，我们负载 MIL-53（Fe）纳米颗粒作为助催化剂以捕获光生空穴，从而促进光生载流子分离。图 7-5（a）中的安培 i-t 曲线清楚地表明，负载 MIL-53（Fe）的 Fe 掺杂的 $BiVO_4$ 的光电流密度为 $1.15mA/cm^2$，远高于掺杂 Fe 离子的 $BiVO_4$（$0.86mA/cm^2$）和 $BiVO_4$（$0.65mA/cm^2$）的光电流密度。此外，图 7-5（b）是在模拟太阳光照射下，电势范围相对于 SCE 为 0.2 至 1.4V 时，$BiVO_4$、Fe 掺杂的 $BiVO_4$ 和 MIL-53（Fe）改性 Fe 掺杂的 $BiVO_4$ 的 LSV，这也可以证明用 MIL-53（Fe）改性 Fe 掺杂的 $BiVO_4$ 在各种偏置电压下显示出最高的光电流密度。

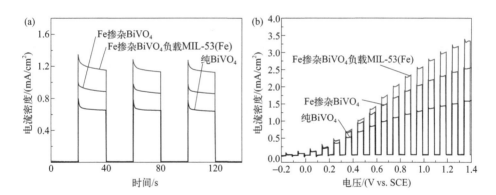

图 7-5　$BiVO_4$、Fe 掺杂的 $BiVO_4$ 和 MIL-53(Fe) 改性 Fe 掺杂的 $BiVO_4$ 的安培 i-t 曲线（a）和线性扫描伏安图（b）

在稳定性实验［图 7-6（a）］中扩大辐照时间，可以清楚地看到 $BiVO_4$ 的光电流将显著降低。为了解决 $BiVO_4$ 的稳定性问题，我们最初提出用微量 Fe 离子取代 Bi 离子。可以观察到，光电流在短时间内稳定，然后在整个 PEC 测量期间保持固定值，这表明 Fe 掺杂大大提高了 $BiVO_4$ 的 PEC 稳定性，这可能是因为用 Fe 离子替代 Bi 离子可以改善 $BiVO_4$ 晶体结构并消除晶体缺陷。根据 MIL-53（Fe）改性的掺杂 Fe 的 $BiVO_4$ 的稳定性测试，可以推断出，如果仅采用 Fe 掺杂，$BiVO_4$ 始终能够表现出高度稳定的 PEC 性能。此外，在系统开路电势下，使用 5mV 的振幅测量了 Fe 掺杂的 BiVO₄ [无论是否沉积 MIL-53（Fe）] 的 EIS，涵盖 0.1～10⁵Hz 的频率范围。Nyquist 曲线中的拱形表示工作电极上的电荷转移动力学，因为半圆的直径反映了电荷转移电阻。如图 7-6（b）所示，MIL-53（Fe）改性的 Fe 掺杂的 BiVO₄ 比 BiVO₄ 和 Fe 掺杂的 BiVO₄ 具有更小的阻抗弧半径，这意味着 MIL-53（Fe）改性的 Fe 掺杂的 BiVO₄ 具有比其他样品更低的电荷转移电阻和更高的电子迁移率。

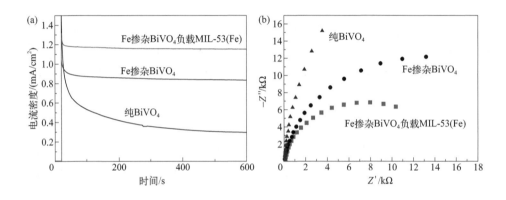

图 7-6　i-t 稳定性测试（a）以及含 MIL-53（Fe）和无 MIL-53（Fe）的 Fe 掺杂的 BiVO₄ 的电化学阻抗谱的 Nyquist 图

7.4　结论

综上所述，在整个 PEC 实验中发现 Fe 离子掺杂可以改善 BiVO₄ 的稳定性和 PEC 性能，这可能是因为 Fe 离子的掺杂可以改善 BiVO₄ 晶体结构并消除晶体缺陷。此外，Fe 掺杂可以与助催化剂协同作用，进一步提高 BiVO₄ 的 PEC 性能。在此，我们设计并制备了 MIL-53（Fe）改性的 Fe 掺杂的 BiVO₄，其表现出比 BiVO₄ 和 Fe 掺杂的 BiVO₄ 高得多的 PEC 性能和稳定性。这是因为 MIL-53（Fe）可以捕获光生空穴，从而提高电荷分离效率并增强 PEC 活性。因此，这可能为将来的异质结构设计和在 PEC 应用中使用助催化剂提供一个备选方案。

参考文献

[1] Hisatomi T, Kubota J, Domen K. Recent advances in semiconductors for photocatalytic and photoelectrochemical water splitting[J]. Chem Soc Rev, 2014, 43: 7520-7535.

[2] Khaselev O, Turner J. A monolithic photovoltaic-photoelectrochemical device for hydrogen production via water splitting[J]. Science, 1998, 280: 425-427.

[3] Reece S, Hamel J, Sung K, et al. Wireless solar water splitting using silicon-based semiconductors and earth-abundant catalysts[J]. Science, 2011, 334: 645-648.

[4] Fujishima A, Honda K. Electrochemical photolysis of water at a semiconductor electrode[J]. Nature, 1972, 238: 37-38.

[5] Hong S, Lee S, Jang J, et al. Heterojunction $BiVO_4/WO_3$ electrodes for enhanced photoactivity of water oxidation[J]. Energy Environ Sci, 2011, 4: 1781-1787.

[6] Jo W, Jang J, Kong K, et al. Phosphate doping into monoclinic $BiVO_4$ for enhanced photoelectrochemical water oxidation activity[J]. Angew Chem Int Ed, 2012, 51: 3147-3151.

[7] Xi G, Ye J. Synthesis of bismuth vanadate nanoplates with exposed {001} facets and enhanced visible-light photocatalytic properties[J]. Chem Commun, 2010, 46: 1893-1895.

[8] McDonald K, Choi K. A new electrochemical synthesis route for a BiOI electrode and its conversion to a highly efficient porous $BiVO_4$ photoanode for solar water oxidation[J]. Energy Environ Sci, 2012, 5: 8553-8557.

[9] Li C, Wang S, Wang T, et al. Monoclinic porous $BiVO_4$ networks decorated by discrete $g-C_3N_4$ nano-islands with tunable coverage for highly efficient photocatalysis[J]. Small, 2014, 10: 2783-2790.

[10] Kim T, Choi K. Nanoporous $BiVO_4$ photoanodes with duallayer oxygen evolution catalysts for solar water splitting[J]. Science, 2014, 343: 990-994.

[11] Rao P, Cai L, Liu C, et al. Simultaneously efficient light absorption and charge separation in $WO_3/BiVO_4$ core/shell nanowire photoanode for photoelectrochemical water oxidation[J]. Nano Lett, 2014, 14: 1099-1105.

[12] Chang X, Wang T, Zhang P, et al. Enhanced surface reaction kinetics and charge separation of p-n heterojunction $Co_3O_4/BiVO_4$ photoanodes[J]. J Am Chem Soc, 2015, 137: 8356-8359.

[13] Kong H, Won D, Kim J, et al. Sulfur-doped $g-C_3N_4/BiVO_4$ composite photocatalyst for water

oxidation under visible light[J]. Chem Mater, 2016, 28: 1318-1324.

[14] Zhong D, Choi S, Gamelin D. Near-complete suppression of surface recombination in solar photoelectrolysis by "Co-Pi" catalyst-modified W: $BiVO_4$[J]. J Am Chem Soc, 2011, 133: 18370-18377.

[15] Pilli S, Furtak T, Brown L, et al. Cobalt-phosphate (Co-Pi) catalyst modified Mo-doped $BiVO_4$ photoelectrodes for solar water oxidation[J]. Energy Environ Sci, 2011, 4: 5028-5034.